Der i-Kosmos
Macht, Mythos und Magie einer Marke

The i-cosmos
Might, myth and magic of a brand

Volker Fischer

Der i-Kosmos
Macht, Mythos und Magie einer Marke

The i-cosmos
Might, myth and magic of a brand

Edition Axel Menges

© 2011 Edition Axel Menges, Stuttgart/London
ISBN 978-3-936681-48-2

Alle Rechte vorbehalten, besonders die der Übersetzung in andere Sprachen.
All rights reserved, especially those of translation into other languages.

Reproduktionen/Reproductions: Christine Ramme, Frankfurt am Main
Druck- und Bindearbeiten/Printing and binding: Graspo CZ, a. s., Zlín, Tschechische Republik/Czech Republic

Übersetzung ins Englische/Translation into English: SATS Translation Services, Katja Steiner, Tengen
Gestaltung/Design: CR DESIGN, Christine Ramme, Frankfurt am Main

Umschlagphoto: IPod-Familie und iPad, Apple, 2010
Cover photo: IPod family and iPad, Apple, 2010

Inhalt / Contents

6	Vorwort
8	Prolog: Der i-Kult. Innovationen, Emotionen, Filiationen
16	Externe Voraussetzungen: Zur Entwicklung digitaler »consumer electronics«
22	Interne Voraussetzungen: Das Entstehen der Applemania
26	Der Kosmos der i-Geräte: Der iPod und seine Parallel-, Peripherie- und Klonprodukte
50	Der Kosmos der i-Geräte: Das iPhone und seine Parallel-, Peripherie- und Klonprodukte
60	Der Kosmos der i-Geräte: Das iPad und seine Parallel-, Peripherie- und Klonprodukte
72	Der i-Kosmos: Das Markenimage, das Marketing, die i-Apps und die Apple Stores
86	Epilog: Ansichten, Einsichten, Aussichten
96	Anmerkungen
98	Nutzer-Erfahrungen
102	Glossar
108	Literatur
112	Abbildungsnachweis

7	Preface
9	Prologue: The i-cult. Innovations, emotions, filiations
17	External preconditions: On the development of digital consumer electronics
23	Internal preconditions: The emergence of applemania
27	The cosmos of the i-devices: The iPod and its peripheries, filiations, and adaptations
51	The cosmos of the i-devices: The iPhone and its peripheries, filiations, and adaptations
61	The cosmos of the i-devices: The iPad and its peripheries, filiations, and adaptations
73	The i-cosmos: The brand image, the marketing, the i-apps and the Apple stores
87	Epilogue: Views, insights, outlooks
97	Notes
99	User experiences
103	Glossary
109	Bibliography
112	Credits

Vorwort

Das vorliegende Buch, welches eine gleichnamige Ausstellung im Museum für Angewandte Kunst in Frankfurt am Main begleitet, ist weltweit die erste umfassende designhistorische Analyse der i-Geräte, ihrer Parallel- und Peripherieprodukte sowie ihrer kommunikationstheoretischen und sozialpsychologischen Implikationen.

Unter rein technikgeschichtlichen Aspekten scheinen der iPod, das iPhone und das iPad durchaus normale Weiterentwicklungen in ihren jeweiligen Gerätegattungen zu sein. Aber offensichtlich hat diese Produktfamilie eine sowohl technische wie ästhetische, vor allem aber benutzerlogistische Qualität erreicht, die ihr eine nobilitierte Alleinstellung sichert.

Die Designgeschichte kann die innovative Kraft dieser Produkte objektiv in historische Entwicklungen einordnen. Die emotionale unmittelbare Estimierung und Faszination, die diese Geräte auslösen, wird damit historisch und technikgeschichtlich verortet. Rund um die Produkte haben sich inzwischen buchstäblich Tausende von Ergänzungs- und Parallelprodukten etabliert. Die Geschwindigkeit der Produktnovitäten degradiert dabei den Nutzer oft zum hoffnungslos hinterherhechelnden Nostalgiker, den beim Neukauf eines Gerätes inzwischen unweigerlich das Gefühl überkommt, ein Auslaufmodell zu erwerben.

Ohne nachvollziehbare Begründung hat sich die deutsche Zentrale des Unternehmens Apple in München erstaunlicherweise geweigert, die Ausstellung und den Katalog zu unterstützen. Ebenso hat die größte deutsche Handelskette für Apple-Produkte Gravis mit Hinweis auf das kalifornische Unternehmen sowie die Zeche Zollverein Essen (Red Dot Design Award) in vorauseilendem Gehorsam eine Kooperation abgelehnt. Wenn bei einer Ausstellung und ihrem Katalog mit mehr als 300 Produkten von gut 130 Unternehmen just dasjenige nicht kooperationsbereit ist, dem die Ausstellung in erster Linie gilt, dann ist das nicht nur ärgerlich, sondern auch ein kulturpolitischer Affront und eine Behinderung der wissenschaftlichen Arbeit einer öffentlichen Kulturinstitution.

Abbildungen und Leihgaben kommen deshalb nun von befreundeten Institutionen, Sammlern, Firmen und Agenturen, denen wir umso mehr danken. Neben der Pinakothek der Moderne und der Neuen Sammlung in München sowie dem Grassi-Museum für Angewandte Kunst Leipzig danken wir dem IF International Forum Design in Hannover, dem Lürzers Archiv in Wien sowie dem Unternehmen re:Store, eine Tochtergesellschaft der russischen ECS Group, für zahlreiche Leihgaben. Auch allen anderen Leihgebern sei an dieser Stelle herzlich gedankt.

Im einzelnen: 1&1 Internet AG, 4tiitoo, Acer, Altec Lansing, Amazon, Apple USA, Apronti, Archos, Arts in the city, Asus, Audi, Avox, Bang & Olufsen, Barnes & Noble, Batoul Apps, Beatles Museum Halle, Belkin, Bergmann, *Bild*, BlackBerry®, BMW, Bowers & Wilkins, Box telecom, Brabus, Cisco, Creative, DeinDesign, Dell, Digital Group Audio, Deutsches Patent- und Markenamt, Sebastian Düvel, Edifier, Elastica, Elecom, John Fader, Finite Elemente, *Focus*, Frankie's Garage, Freiwild, frog design, Garmin, Geneva, Golfoholic, Google, Grove Made, Grundig, hama, Hanvon, Hottrix, HTC, Iriver, Jailbreak collective, James Law Cybertecture International, JBL, Junaio, Jung, Kardon Harman, David Jon Kassan, Kenwood, Klein & More, Koziol, Dirk Kunde, Kunstflug, Regis Laborie, Lapàporter, Lenco, Lenovo, LG, Libratone, Peter Morgan, Markengold, Maxenia, Metaio, Microsoft, Motorola, Msi Computers, Native Union, Nike, Nintendo, Nokia, Odys, Pack and smooch, Palm, Panasonic, Parrot, Philips, Pioneer, Playbutton, proidee, Research In Motion®, Rotaliana, Samsung, SanDisk, Scoopertino, Senotech, Sharp, Silver Seiko, Sonoro, Sonos, Sony, Sony Ericsson, Soulra, *Der Spiegel*, Jeff Stieler, Switcheasy, Markus Tacker, Tchibo, Technabob, Text 100, Thalia Buchhandelsgruppe, Tivoli, Toshiba, Trademark PR, TrekStor, Twelve South, Urban Tool, ViewSonic, Xayni und Zumtobel.

Der Kunstgewerbeverein als Förderverein des Museums für Angewandte Kunst Frankfurt hat dankenswerterweise ebenfalls die Ausstellung unterstützt. Ein besonderer Dank geht an Christine Ramme, die nicht nur gravierend an der Objekt- und Bildrecherche beteiligt war, sondern auch das Layout des Kataloges gestaltet hat.

Ulrich Schneider
Direktor des Museums für Angewandte Kunst Frankfurt

Volker Fischer
Oberkustos und Kurator für Design des Museums für Angewandte Kunst Frankfurt

Preface

This book, which accompanies an exhibition with the same title at the Museum für Angewandte Kunst in Frankfurt am Main, is the first of its kind. It is a comprehensive design-historical analysis of i-devices, parallel and peripheral products, as well as the implications with regard to communication theory and social psychology.

Seen purely under the aspects of the history of technology, the iPod, the iPhone and the iPad seem to be absolutely normal developments in their pertaining device genres. But this product family has obviously achieved a technological and aesthetic quality, but mainly a user-logistics quality, that secures it an ennobled and unique status.

Design history can objectively classify the innovative power of these products according to historical developments. The immediate emotional assessment and fascination triggered by these devices is thus localized in terms of history and the history of technology. Today, literally thousands of peripheral and parallel products have sprung up around the i-products. The speed of product innovation often degrades users into hopeless sentimentalists who are always lagging behind, who are inevitably overcome by the feeling that they are purchasing an obsolete model when they buy a device.

Without understandable rationale, the German head office of Apple, surprisingly, has refused to support the exhibition and catalogue. The biggest German trade chain for Apple products, Gravis, also rejected cooperation in anticipatory obedience, referring to the California company as well as Zeche Zollverein Essen (Red Dot Design Award). If the company to which an exhibition and catalogue with more than 300 products from almost 130 companies is first and foremost dedicated is unwilling to cooperate, then that is not only irritating but also a cultural-political affront and a hindrance to the scientific work in a public cultural institution.

For this reason, images and loans now come from friendly institutions, collectors, the trade and agencies; we would like to thank them for their cooperation. Alongside the Pinakothek der Moderne and the Neue Sammlung in Munich as well as the Grassi-Museum für Angewandte Kunst Leipzig, we would like to thank the IF International Forum Design in Hanover, the Lürzers Archiv in Vienna, and the company re:Store, a subsidiary of the Russian ECS Group, for numerous loans. Our thanks also go to all the other lenders.

They are: 1&1 Internet AG, 4tiitoo, Acer, Altec Lansing, Amazon, Apple USA, Apronti, Archos, Arts in the city, Asus, Audi, Avox, Bang & Olufsen, Barnes & Noble, Batoul Apps, Beatles Museum Halle, Belkin, Bergmann, *Bild*, BlackBerry®, BMW, Bowers & Wilkins, Box telecom, Brabus, Cisco, Creative, DeinDesign, Dell, Digital Group Audio, Deutsches Patent- und Markenamt, Sebastian Düvel, Edifier, Elastica, Elecom, John Fader, Finite Elemente, *Focus*, Frankie's Garage, Freiwild, frog design, Garmin, Geneva, Golfoholic, Google, Grove Made, Grundig, hama, Hanvon, Hottrix, HTC, Iriver, Jailbreak collective, James Law Cybertecture International, JBL, Junaio, Jung, Kardon Harman, David Jon Kassan, Kenwood, Klein & More, Koziol, Dirk Kunde, Kunstflug, Regis Laborie, Lapàporter, Lenco, Lenovo, LG, Libratone, Peter Morgan, Markengold, Maxenia, Metaio, Microsoft, Motorola, Msi Computers, Native Union, Nike, Nintendo, Nokia, Odys, Pack and smooch, Palm, Panasonic, Parrot, Philips, Pioneer, Playbutton, proidee, Research In Motion®, Rotaliana, Samsung, SanDisk, Scoopertino, Senotech, Sharp, Silver Seiko, Sonoro, Sonos, Sony, Sony Ericsson, Soulra, *Der Spiegel*, Jeff Stieler, Switcheasy, Markus Tacker, Tchibo, Technabob, Text 100, Thalia Buchhandelsgruppe, Tivoli, Toshiba, Trademark PR, TrekStor, Twelve South, Urban Tool, ViewSonic, Xayni, and Zumtobel.

The Kunstgewerbeverein (Handicrafts Association) as the supporting organization of the Museum für Angewandte Kunst Frankfurt kindly sponsored the exhibition as well. Special thanks go to Christine Ramme, who not only played a decisive role in the object and image research but also designed the catalogue's layout.

Ulrich Schneider
Director of the Museum für Angewandte Kunst Frankfurt

Volker Fischer
Senior curator of the design department of the Museum für Angewandte Kunst Frankfurt

Prolog: Der i-Kult. Innovationen, Emotionen, Filiationen

Der biblische »Apfel der Erkenntnis« hat seit über drei Jahrzehnten einen elektronischen Nachfolger, der die Symbolik der »Vertreibung aus dem Paradies« ins Gegenteil einer »Teilhabe am Paradies« umgekehrt hat. Dieser Nachfolger ist das Logo des amerikanischen Unternehmens Apple, welches für eine der erstaunlichsten Erfolgsgeschichten der Industriegeschichte steht. Allein der Anblick dieses angebissenen Apfels löst euphorische Begehrlichkeiten aus. Dabei erscheint der angebissene Logo-Apfel – immer schon Symbol für eine quasireligiöse Überzeugungsgemeinschaft – inzwischen auch den zupackenden Biß des Unternehmens auf Kunden und Konkurrenten ebenso wie die Kreation neuer Bedürfnisse zu symbolisieren: Insofern kann man durchaus von einer Applemania sprechen. Diese Applemania hat durchaus den Charakter einer profanierten Gnosis: Die Dinge sind nicht nur, was sie sind, sondern auch, was sie scheinen und versprechen. Wie in den neutestamentlichen Schriften vereint die Apple-Gemeinde eine vertiefte Glaubenseinsicht und -gewißheit in geoffenbarte Wahrheiten. Die solchermaßen den Apple-Geräten zuerkannten numinosen Qualitäten machen sie gewissermaßen zu Techno Fetischen, ja Techno-Reliquien.

Schauen wir zunächst auf die Entwicklung des Firmenlogos, denn dieses ist immer ein Hinweis auf das Selbstverständnis eines Unternehmens sowie darauf, wie es von außen gesehen werden will. Daniel Haas bemerkt zum Apple-Logo: »Weltruhm entsteht, wenn die natürliche Form zur Kontur gerinnt. Globale Bedeutung kommt in der Übersetzung von kreatürlichen in symbolischen Ausdruck zustande. Es scheint, als brauchte man den Apfel nur, damit er – in graphischer Gestalt, mit herausgebissener Ecke – den weltweiten Siegeszug einer technologischen Idee anzeigt (und darüber hinaus die Durchsetzung eines Sets lebensweltlicher Haltungen, kurz: eines Lifestyles).«[1]

Das erste Logo im Stil eines barocken Kupferstichs, 1976 entworfen von Steve Jobs und Ron Wayne, zeigt Isaac Newton, unter einem Apfelbaum sitzend. Der Apfel am Zweig über seinem Kopf – dessen Fallen ihn der Legende nach zu seinem 1666 formulierten berühmten Gravitationsgesetz inspirierte – ist wie eine Reliquie durch eine Gloriole hervorgehoben. Eine flatternde Banderole zeigt die Wortmarke »Apple Computer Co.« Graphisch wirkt dieses Schwarzweiß-Logo wie eine Hommage an die amerikanische Pionierzeit und zierte auch das »Operation Manual« des Apple II. Offenbar war dieses Logo drucktechnisch für Verkleinerungen ungeeignet und darüber hinaus dann doch zu konservativ, denn nach ein paar Monaten bereits gestaltete Regis McKenna eine schwarze Apfelsilhouette mit Biß als Apple-Logo. Da »beißen« im Englischen »to bite« heißt, stellt sich auch die Konnotation »Byte« ein. Die sieben Farbstreifen des »Regenbogen-Apfels von Rob Janoff, 1977 mit dem Apple II eingeführt, paraphrasieren Paul Rands berühmtes IBM-Logo ebenso, wie der frühe Apple-Slogan »Think different« auf den IBM-Slogan »Think«reagierte. Die erste, mit dem bunten Apfel korrespondierende Wortmarke in Kleinschreibung zeigte leicht verpopte Buchstaben der Letraset-Schrift Motter Tektura. Ab 1980 kam eine Garamond-Schrift zum Einsatz und ab 2002 mit dem Start des eMacs die serifenlose Myriad-Schrift von Adobe, allerdings als spezielle Myriad-Apple-Version. Mit der Markteinführung des semitransparenten iMac 1998 wurde das Logo wieder einfarbig in wechselnden Farben denen der Computer angepaßt und leicht plastisch gestaltet.

Dieses Buch will keine erschöpfende Geschichte des Unternehmens Apple Inc. abhandeln, die zudem bereits mehrfach erzählt worden ist. Vielmehr konzentriert es sich auf die unterhaltungselektronischen Produkte der i-Familie, also auf die iPods, iPhones und iPads. Die wechselvolle Geschichte der Personal Computer von Apple wird nur zusammenfassend als Kontext erläutert, obwohl auch hier die generelle Tendenz der Konvergenz zwischen PC-Welten und Unterhaltungselektronik evident ist.

Vor allem aber bilden der iPod, das iPhone und das iPad eine Trias, die den Verständnis- und Erfahrungshorizont der entsprechenden Produktgattungen folgenreich verändert hat. Es geht um neue, vernetzte Nutzungen, um bisher so nicht gekannte Symbiosen von Hard- und Software und um ein geradezu kultisches Verhältnis zu unterhaltungselektronischen Geräten. Die Käufer der i-Produkte finden sie »sexy«. Wenn man unbedingt etwas haben will, was man eigentlich nicht braucht, ist der Fokus vom Nutzen auf die Begehrlichkeit verlagert und eine Balance erreicht zwischen »must-have« und »nice-to-have«. Mit einem Wort des spanischen Autors Mario Perniola: Es gibt auch einen anorganischen »Sex-Appeal«. Konsumforscher sehen in Apple insofern weniger einen Technologielieferanten als vielmehr einen »Erlebnisprovider«, der dominant den Lebensstil der heutigen und der nächsten Generation definiere. Selbst Automanager renommierter und teurer Luxusmarken zeigen sich irritiert darüber, daß jüngere Kunden zuerst danach fragen, ob ihr iPod oder iPhone in das neue Modell passen würden. Und so wollen auch weit größere Firmen plötzlich so werden wie Apple: schick, begehrt und profitabel.

Man muß nicht so weit gehen wie manche Objektfetischisten, die gegenüber Laptops, Dampfmaschinen oder Hochhäusern libidinöse Gefühle entwickeln.[2] Wohl aber bleibt der Hype auf neue Apple-Produkte bemerkenswert, der in einer Art historischer Koinzidenz die Reaktionen auf die gut vierzig Jahre vorher produzierten Schallplatten der Beatles entsprechen, die als sog. Beatlemania (vom Herbst 1963 bis zum Sommer 1966) in die Pop-Geschichte eingingen. Deren Produktionsfirma hieß ab dem Frühjahr 1968 Apple Records, und es ist kein Zufall, daß der Hersteller der i-Geräte heute genauso heißt. Der Gründer von Apple Inc. in Kalifornien, Steve Jobs, bestätigte, von seiner Lieblingsband zum Namen »Apple« inspiriert worden zu sein. Außerdem ernährte er sich damals

Prologue: The i-cult. Innovations, emotions, filiations

For more than three decades the biblical »apple of knowledge« has had an electronic successor, which has reversed the symbolism from »expulsion from paradise« to »participation in paradise«. This successor is the logo of the American company Apple, which has had one of the most amazing success stories in industrial history. A look at this bitten into apple alone – always a symbol of a quasi-religious community of faith – causes euphoric, covetous desires, although the bitten-into apple logo also seems to symbolize the proactive bite of the company at customers and competitors as well as the creation of new needs: in this regard, one may absolutely speak of applemania. This applemania absolutely has the character of a profanized gnosis: things are not only what they are but also what they seem to be and promise. Like in the scripts of the New Testament the Apple community is united by a deepened insight and certainty of faith in revealed truths. The numinous qualities that are awarded to the Apple devices in a sense make them techno fetishes, even techno relics.

Let us first look at the development of the corporate logo because it is always an indication of self-image and of how a company wishes to be perceived from the outside. Daniel Haas remarked about the Apple logo: »World fame comes about when the natural form coagulates into a contour. Global importance emerges from the translation of natural expression into symbolic expression. It seems as though it only takes the Apple in order to indicate – in a graphic form, with a bitten-out corner – the triumphant global conquest of a technological idea (and beyond that, the enforcement of a set of everyday life-world attitudes, in short: a lifestyle).«[1]

The first logo, designed in the style of a baroque copper etching by Steve Jobs and Ron Wayne in 1976, shows Isaac Newton sitting under an apple tree. The apple on a branch above his head – according to legend, a falling apple inspired him to formulate his famous law of gravity in 1666 – is emphasized like a relic with a halo. A flattering banderole bears the inscription »Apple Computer Co«. Graphically, this black-and-white logo looks like an homage to the American pioneering days and also adorned the »Operation Manual« of the Apple II. Obviously, however, the logo was not suitable for downscaling with regard to print technology and, in addition, it was too conservative because only a few months later Reggis McKenna designed a black silhouette of an apple with a bite taken out as the Apple logo. The word »bite« also includes the connotation of »byte.« The seven colored stripes of the »rainbow apple« by Rob Janoff, introduced with the Apple II in 1977, paraphrases Paul Rand's famous IBM logo just as the early Apple slogan »Think different« was a response to IBM's slogan »Think.« The first word mark corresponding with the colorful apple in small letters showed slightly pop-style characters of the Letraset font Motter Tektura. Starting in 1980, a Garamond font was used and, starting in 2002, Adobe's sans-serif Myriad font accompanied the launch of the eMac, but as a special Myriad Apple version. With the market introduction of the semi-transparent iMac in 1998 the logo again became monochromatic in various colors that matched the computers and had a slightly sculptural design.

This book does not aim to offer an exhausting history of the company Apple Inc., which has been repeatedly told. It rather focuses on the entertainment electronics products of the i-family, the iPods, iPhones, and iPads. The varied history of Apple's personal computers is explained as a summarizing context, although here too the general tendency of the convergence between PC worlds and entertainment electronics is evident.

Above all, however, the iPod, the iPhone and the iPad form a triad that has profoundly changed the associated product genres' horizon of understanding and experience. It is about new networked usages, thus far unknown symbioses of hardware and software, and an almost cult-like relationship with electronic entertainment devices. The buyers of the i-products consider them »sexy.« If one absolutely has to have something one does not really need, the focus on utility is shifted and the object instead becomes a covetous possession and a balance between »must-have« and »would-

1. Steve Jobs, Ron Wayne: Logo »Apple Computer Co.«, 1976.
2. Reggis McKenna: first Apple Logo with type »Motter Tektura« by Othmar Motter, 1976.
3. Logo with type »Apple Myriad« by Robert Slimbach and Carol Twombly, since 2002.

angeblich nur von Obst, unter anderem eben von vielen Äpfeln. Die Namensgleichheit hat seit drei, vier Jahrzehnten immer mal wieder zu launigen gerichtlichen Auseinandersetzungen von epischer Länge geführt, die die Rechtsanwälte der Beatles mit immer weniger Glanz gewannen, bis sie 2006 den letzten Prozeß schließlich verloren. »Dieser Streit ist ein Grund, warum es die Musik der Beatles bis heute nicht legal im Internet zum Download gibt, Apple ist mit seinem iTunes Store marktbeherrschend. Aus den Ankündigungen, eine eigene Download-Plattform zu erstellen, ist bis heute nichts geworden. Intern sollen bei Apple (Records) längst Studien liegen, die nachweisen, wie viel Geld der Konzern durch die Weigerung, ins Netz zu gehen, verloren hat.«[3] In einer Art Buy out übernahm im Februar 2007 die kalifornische Firma alle Rechte am Apfelnamen und Apfellogo von Apple Records. Seitdem benutzt die Beatles-Firma dieses Warenzeichen in Lizenz und muß dies dem Computerunternehmen bezahlen. Über die verhandelten Summen wurde Stillschweigen vereinbart. Heute gibt es die Beatles-Alben durchaus zum Downloaden bei verschiedenen Portalen, z. B. Amazon, aber eben nicht im iStore von Apple. Seit Ende 2010 allerdings gibt es alle Alben der Beatles im iTunes Store. Und ein USB-Stick in Apfelgrün hat als Verpackung einen dreidimensionalen, metallisch grün schimmernden Apfel mit dem Beatles-Logo, in die man den Stick einschiebt.

So wie es die erwähnte Beatlemania gab, die darin bestand, daß junge Mädchen beim Anblick der Fab Four reihenweise in hysterische Verzückungen gerieten, so geraten die Nerds bei jeder Neuvorstellung eines neuen Apple-Gerätes in unkontrollierbare Kaufräusche. Bedenkt man, daß die Beatles-Firma Apple Records heißt, dann bekommt die Parallele zwischen Beatlemania und Applemania eine geradezu teleologische Perspektive.

Der Buchstabe »i« als Produktlinienkennzeichnung ist besonders im Englischen ein idealer Assoziationsauslöser. So stehen Konnotationen an »internet«, »information«, »intelligence« und »interface« für den Wissensaspekt der i-Geräte, »identity«, »individual«, »inspiration« und »innovation«, vielleicht auch »integrity« für die emotionale Seite. Und nicht zuletzt steht das »i« auch für »infotainment«. Außerdem bedeutet dieser Buchstabe im Englischen auch »ich« und verweist damit auf die affektive Bindung zwischen Gerät und Benutzer. Eine Affinität zur Mehrdeutigkeit der Sprache zeichnet im Kontext des Unternehmens auch das Wort Apps aus, Natürlich ist es die Abkürzung des Begriffs »application« als Bezeichnung für die Anwendungsprogramme der i-Geräte. Gleichzeitig aber mag man Apps auch als Kurzform von »Apple« verstehen. Und jeder Apple Store ist ja auch ein App Store, obwohl diese eher im Internet frequentiert werden. Zudem haben die i-Geräte die Charakteristik ihre Produktgattungen offenbar jeweils so gravierend verändert, daß die meisten Hersteller von MP3-Geräten, Mobiltelephonen und Tablet-Computern sich inzwischen an den Apple-Produkten orientieren. Designtheoretisch gesprochen, handelt es sich bei diesen i-Geräten jeweils um komplette Paradigmenwechsel innerhalb der drei Produktgattungen.

1995 bezeichneten Joseph Bower und Clayton Christensen von der Harvard Business School solche Entwicklungen, die einen radikalen Bruch mit einer vorherrschenden Technik darstellen, als »disruptive Technologien«. Sie führen stets dazu, daß die Karten auf den damit verbundenen Märkten neu gemischt werden. Musterbeispiele sind die inzwischen nahezu vollständige Verdrängung der klassischen Photographie auf Film (seit 1925) durch die Digitalkameras (seit 1991), jene der Tonbandspule (seit 1935) durch die Audiokassette (seit 1963), diese durch die CD bzw. Mini-CD (seit 1981), das Super 8-Filmformat (seit 1965) durch die Videokassette (seit 1976) und diese durch die DVD (seit 1997). Die CD-Rom (seit 1979) wird gegenwärtig durch das Speichermedium der USB-Sticks (seit 2000) abgelöst. Die Diskette (seit 1969) wurde durch die Floppydisc (seit 1976) abgelöst, deren Produktion allerdings im März 2001 eingestellt wurde, so daß die Diskette sie überlebte. Aber auch diese beiden Speicherformate sind praktisch verschwunden. Die i-Geräte sind ein vorläufiger Höhepunkt solcher Verdrängungen, weil bei ihnen nicht nur eine Speichertechnologie eine ältere ersetzt, sondern weil im Grunde alle bisherigen Aufzeichnungs- und digitalen Kommunikationsmedien – Photographien, Photo- und Videodateien, Musikdateien, Info- und Nachrichtendienste, mobiles Internet, telephonieren, SMS und Computerspiele – durch eines der i-Geräte jeweils fast vollständig ersetzt werden kann. So ersetzt ein Gerät acht oder zehn andere, den gan-

be-nice-to-have« is achieved. To quote the Spanish author Mario Pernola: an inorganic »sex appeal« exists as well. Consumer researchers in this sense consider Apple to be more an »experience provider« that dominantly defines the lifestyle of today's generation and the next than a technology supplier. Even managers of renowned and expensive luxury automobile brands are irritated by the fact that the first question younger customers ask them is whether the new car will accommodate their iPods or iPhones. And thus even much bigger companies suddenly strive towards becoming like Apple: chic, coveted, and profitable.

One does not have to go as far as some object fetishists, who develop libidinous emotions towards laptops, steam engines or high-rise buildings.[2] Still, the hype over new Apple products remains remarkable; in a kind of historical coincidence, the reactions to the Beatles albums produced almost forty years earlier, which entered pop history as Beatlemania (from fall 1963 until the summer of 1966), correspond to it. Their production company, Apple Records, was established in the spring 1968, and it is no coincidence that the manufacturer of the i-devices is called Apple today. The founder of Apple Inc. in California, the Steve Jobs, confirmed that his favorite band inspired the name »Apple.« In addition, he supposedly ate only fruit and, at the time, many apples. This similarity of names has repeatedly led to witty court confrontations of epic length over the past three or four decades, which the lawyers of the Beatles won with increasingly less glamour before losing the final trial in 2006. »This dispute is one reason the music of the Beatles is not legally available for download on the Internet to this day; Apple dominates the market with its iTunes store. The announcements that a special download platform would be established have not born fruit so far. Internally, Apple (Records) is said to have conducted studies that reveal the amount of money that the group has lost due to its refusal to go online for such a long time.«[3] In a kind of buyout, the California company took over the rights to the apple name and apple logo of Apple Records in February 2007. Since then, the Beatles company has been licensing the trademark from Apple Inc. and has to pay the computer company for it. Both sides agreed to not disclose the negotiated amounts. Today, the Beatles albums are available for download on various portals, for example Amazon, but not in Apple's iStore. Since the end of 2010, however, all Beatles albums have been available at the iTunes store. And a USB memory stick in apple-green has a shimmering, metallic green, three-dimensional apple with the Beatles logo on the packaging into which the stick is inserted.

Just as the aforementioned Beatlemania existed and caused hysterical rapture among throngs of young girls when seeing the Fab Four, computer nerds go on uncontrollable shopping frenzies each time a new Apple device is introduced. If one considers that the Beatles company was called Apple Records, an almost teleological perspective opens for the similarities between Beatlemania and applemania.

The letter »i« as a product line identification is an ideal trigger for associations in English. Connotations of »internet«, »information«, »intelligence« and »interface« stand for the knowledge aspect of the three i-devices; »identity«, »individual«, »inspiration« and »innovation«, perhaps also »integrity«, represent the emotional side. And, not least, the »i« also stands for »infotainment.« Naturally, the »I« in English refers to the ego and hints at the affective connection between the device and its user. The word app is also characterized by an affinity to the ambiguity of language in the context of the company. Naturally, it is the abbreviation of the term »application« to describe the application programs for the i-devices. At the same time, however, app can be understood as a short form of »Apple.« And every Apple store is also an app store, although most apps are obtained through the app store on the Internet.

In addition, each i-device has changed the characteristics of its respective product genre to such a profound degree that most manufacturers of MP3 devices, mobile phones and desktop computers now look to Apple's products for orientation. Speaking in terms of design theory, these i-devices are each complete paradigm changes within their product genres.

In 1995, Joseph Bower and Clayton Christensen of the Harvard Business School described such developments that represent a radical break with a prevailing technology as »disruptive technolo-

4. Rob Janoff: 2nd Apple Logo with type »Apple Garamond« by Tony Stan, 1977, compaign »Think different«, 1997.
5. Rob Janoff: 2nd Apple Logo with type »Apple Garamond« by Tony Stan, 1950, on Macintosh PowerBook 140, 1991.
6. Logo AirPort Base Station, Apple, 1999.
7. Logo Personal Computer iMac, Apple, 1995.

zen Dschungel der »consumer electronics«. Bei Licht besehen, kann man dies im Sinne einer Produkt-Vermeidung durchaus ökologisch nennen.

Für einen heute Dreißigjährigen hat es also zehn, zwölf verschiedene Speichermedien gegeben, und hier sind nur die wichtigsten genannt, wohlgemerkt in der Hardware; permanente Updates in den Software-Datenträgern kommen noch hinzu. Wir alle sind ebenso Auslöser wie Betroffene eines Techno-Darwinismus, der dem Prinzip »struggle for life« auch auf der Ebene der Platinen und Chips zur Geltung verhilft. Insofern sind in den Kommunikationstechnologien disruptive Brüche die Regel. Heute aber sind wir mit einem Verständniswandel von Kommunikation überhaupt konfrontiert. Sie ist in einer Weise dezentral, entortet und mobil wie noch niemals in der Geschichte. Dabei kommunizieren wir mit Geräten ebenso wie über interaktive Geräte mit anderen Menschen. Die affektiven Bindungen zwischenmenschlicher Kommunikation entstehen so oft auch gegenüber Geräten.

Im Folgenden werden wir die erwähnte produktkulturelle Paradigmatik der i-Welten aufblättern, Parallel- und Konkurrenzprodukte zeigen, ihre jeweiligen gattungsspezifischen Voraussetzungen erläutern, ihre Vermarktungs- und Copyright-Strategien beleuchten und deutlich machen, warum eine weltweite Benutzergemeinde geradezu »i-phorisch« einer »i-philie« frönt.

Unter semiotischen Gesichtspunkten kann man die iPhones und iPads und ihre Distribution im Sinne der von Roland Barthes beschriebenen »Mythen des Alltags«[4] als ein strategisches Mystifikationsverfahren beschreiben, das zum Ziel hat, das Publikum mit einer Prise des Mirakulösen zu impfen, um es dann in ein immunisiertes technopragmatisches Befinden zu versetzen. Der aus rituellen Bewegungen entstehende Tanz der Finger auf den gläsernen Displays wirkt wie eine Maske aus Gesten, die eine Handhabbarkeit einfordert, wo doch längst alle Haptik der Knöpfe, Schalter und Regler verschwunden ist. Dieses rituelle Bannen einer nicht mehr vorhandenen Haptik des Gebrauchs verleiht ihr noch, und gerade in ihrer Verneinung, um so größere Bedeutung und Faszination. Konnte Otl Aicher im Kontext von Türgriffen noch statuieren: »Das Greifen ist das Begriffene«[5], so müssen wir beim iPhone und iPad wohl sagen: »Das Tasten ist der Test« oder auch »Das Berühren ist das Berührende«. Man wird kaum ähnliche magische Gebrauchskontexte in der Gegenwart finden wie beim Benutzen dieser Geräte. Die Technik bekommt erneut numinose Komponenten: Hier marginalisiert und entkörperlicht sich die Pragmatik selbst. Dies ist ein durchaus subtiler Vorgang, denn er mystifiziert pragmatische Handhabung zum übersinnlichen Handling. Mit einem Fingertipp die »Welt zu Gast« zu haben, ist nicht so weit weg von Omnipotenz-Phantasien. Mit Daumen und Zeigefinger sich die Welt »aufzuziehen«, heißt nichts anderes, als Gott zu spielen. So wird die Technik zur Metaphysik und die Metaphysik zur Technik. Immanuel Kant läßt grüßen.

Dabei verwischt sich bei den i-Geräten der Unterschied zwischen dem Exzeptionellen und dem Normalen: Sie reklamieren einen Avantgarde-Status und kokettieren gleichzeitig damit, Mainstream zu sein.

Die technikgeschichtlichen Voraussetzungen, die zugleich auch immer gesellschaftliche Veränderungen jenseits der Willensbildung und des Einflusses einzelner Menschen bewirken, hat Peter Sloterdijk beschrieben: »Die durchschnittliche Medienausstattung des späten 20. Jahrhundert-Apartments erlaubt es dem Einzelnen, sich auf einer eigenen virtuellen Insel zu etablieren und Kommunikation von Insel zu Insel zu betreiben. Man vergißt zu oft, daß die Bedingung der modernen Telekommunikation eine Weltform ist, in der die große Mehrheit der Menschen sich wie von Natur aus als Insulaner verstehen. Die Lebensformen des späten 20. Jahrhunderts gleichen einer Hyper-Robinsonade.«[6]

Zweifellos ist die mobile vernetzte Kommunikation eine der dominanten Signaturen unserer Epoche. Nicht nur allzeit erreichbar zu sein, sondern auch alle und alles zu »erreichen«, scheint heute für viele die conditio sine qua non ihrer gesellschaftlichen Existenz und Selbstvergewisserung zu sein. Offline zu sein erscheint in einer solchen Perspektive gleichsam gleichzeitig als innere und äußere Emigration.[7] Wer einmal durch drei, vier Waggons eines ICEs zum Bordrestaurant gegangen ist, der wird den fahlblauen Schimmer von Dutzenden auf den Knien balancierter Laptops nicht vergessen. Eine offensichtlich völlig introvertierte Jungmanager-Kaste ist ersichtlich nicht mehr willens oder in der Lage, für ein paar Stunden – ob zwischen Hamburg und München oder zwischen Frankfurt und Paris – einmal offline zu sein. »Kontaktbörsen«-Apps wie YouTube, Myspace oder Facebook haben längst die Rendezvous unter Uhrtürmchen, die amourösen Billetts und Tanzstundenverabredungen verdrängt. Zwar gibt es noch Pop-Konzerte und Musiksäle; aber immer und überall – vom U-Bahn-Abteil bis in die Toilette – die Lieblingsmusik über die Kopfhörer eines iPod nano zu hören, erscheint den Zwölf- bis Dreißigjährigen wesentlich attraktiver. Man macht sich kaum einen Begriff davon, in welchem Umfang das Alltagsleben und -erleben von Millionen von Menschen heute durch mobile Vermittlungsformen der Unterhaltung, Information und des Wissens geprägt ist. Stationäre Vermittlungsinstanzen wie Bibliotheken, Kinos, Theater, Konzertsäle oder Universitäten werden zusehends marginalisiert: Wir gehen nicht mehr zu den Informationen, sondern die Informationen kommen zu uns. Sie sind frei flottierend, mobil, ubiquitär und zu allen Zeiten und an allen Orten verfügbar geworden. Das erzeugt eine neue Form von Intimität, die aber (als Kehrseite der Medaille) von »Datenkraken« wie Google, Microsoft, zunehmend auch Apple, kontrolliert, überwacht und zu eigenen Zwecken genutzt wird. Wobei diese Überwachung wohl eher eine Konsumentenmassage im Sinne Aldous Huxleys als eine brutale Unterdrückung im Sinne George Orwells ist.

Schon 1986 hatte Neil Postman in seinem Buch *Wir amüsieren uns zu Tode* mit dem schlagenden Untertitel *Urteilsbildung im Zeitalter der Unterhaltungsindustrie* angesichts des Überflusses an Unterhaltungsangeboten diese Urteilsbildung vor den Hintergrund dieser zwei utopischen Szenarien ge-

8. The Beatles: *Hey Jude – Revolution*, Apple Records, 1968.
9. The Beatles: *The Long And Winding Road – For You Blue*, Apple Records, 1970.
10. The Beatles: *Something – Come Together*, Apple Records, 1969.

gies.« They always lead to the cards being reshuffled in the associated markets. Textbook examples include the almost complete displacement of classic film photography (since 1925) by digital cameras (since 1991), reel-to-reel recording tapes (since 1935) by cassettes (since 1963), cassettes by CDs or mini-CDs (since 1981), Super 8 film (since 1965) by video tapes (since 1976) and then DVDs (since 1997). The CD ROM (since 1979) is currently being replaced by the USB memory stick (since 2000). The floppy disk (since 1969) was replaced by the 3-inch floppy diskette (since 1976); however, production ceased in March 2001 and the diskette survived it. But these two storage formats have since disappeared from the computer landscape, as well. The i-devices are a preliminary climax of such replacements because in their case not only does one storage technology replace an older one but because fundamentally all past storage and digital communication media – photographs, photo and video files, music files, information and news services, mobile Internet, telephony, SMS, and computer games – can be almost completely replaced by one of the i-devices. One of them replaces eight or ten others, the entire jungle of consumer electronics. Seen in the proper light, we could absolutely call it ecological in the sense of product avoidance.

Today, a 30 year old has seen ten to twelve different types of storage media, and we have only mentioned the most important ones here, mind you in hardware; constant software updates can be added to that. We are all both catalysts of and parties to a techno-Darwinism; we help get the principle »survival of the fittest« accepted on the level of printed circuit boards and chips as well. In this regard, disruptive breaks are the rule in communication technologies today. However, we are confronted with a transformation of the understanding of communication in general today. In a way, it is more decentralized, delocalized and mobile than ever before. We communicate with other people through passive devices as well as interactive devices. The affective bonds of interpersonal communication thus often also emerge towards devices.

Below, we will browse through the aforementioned product-cultural paradigmatics of the i-worlds, show the parallel and competing products, explain their genre-specific premises, shed light on their marketing and copyright strategies, and clarify why a global user community is almost »i-phorically« indulging in an »i-philia.«

From a semiotic point of view the iPhones and iPads and their distribution can be described as a strategic process of mystification in the sense of the »Mythologies«[4] described by Roland Barthes, which aims to inoculate the public with a pinch of the miraculous in order to then put people in an immunized techno-pragmatical state. The dance of fingers on the glassy displays emerging from ritual movements seems like a mask of gestures that calls for handling, although all of the haptics of buttons, switches and controllers have long been eliminated. This ritualistic spellbinding effect of the no longer existing haptics of use still, and especially in its negation, provides it with even greater importance and fascination. While Otl Aicher, in the context of door handles, was still able to state: »Gripping is what is grasped«[5], we have to say with respect to the iPhone and iPad: »Feeling is the test« or »Touching is what is touching.« It will be hard to find magical contexts of usage in the present that are similar to the use of these devices. Technology again gets numinous components: here, pragmatism marginalizes and disembodies itself. This is an absolutely subtle process because it mystifies pragmatic handling and turns it into extrasensory handling. Having the »world as a guest« at one's fingertips is not too far removed from fantasies of omnipotence. Being able to »raise« the world with one's thumb and index finger means nothing less than playing God. Technology thus becomes metaphysics, and metaphysics technology. Immanuel Kant sends his regards.

In the case of i-devices, the difference between the exceptional and the normal blurs: they claim an avant-garde status and simultaneously flirt with being mainstream.

Peter Sloterdijk described the technological-historical preconditions that simultaneously always cause social changes beyond the shaping of will and the influence of individuals: »The media equipment found in a typical late 20th century home allows individuals to establish themselves on their own virtual island and communicate from one island to the next. Often, one forgets that the requirement of modern telecommunications is a kind of world in which a large majority of people understand themselves as islanders. The way of life in the late 20th century resembles a hyper Robinsonade.«[6]

Without a doubt, networked mobile communication is one of our epoch's dominant characteristics. Being available at any time but also being able to »reach« anyone and everything today seems to be a *conditio sine qua non* for many people with respect to their social existence and self-assurance. In such a perspective, being offline seems to be simultaneously an inner and outer emigration.[7] Anyone who has ever walked through three or four cars of an ICE train to get to the on-board restaurant has certainly noticed the pale blue shimmer of dozens of laptops balanced on the knees of travelers. It seems that an obviously completely introverted class of young managers is no longer willing to be or capable of being offline for even a few hours today, whether between Hamburg and Munich or Frankfurt and Paris. »Contact exchange« apps such as MySpace or Facebook have long displaced the rendezvous under a small clock tower, the romantic tickets to a show or a dance lesson date. There may still be pop concerts and music halls, but listening to one's favorite music anytime and anyplace via headphones and an iPod nano seems to be much more attractive to twelve to thirty year-olds. It is hard to imagine to what extent the day-to-day life and experience of millions of people is influenced by mobile forms of communication, entertainment, information, and knowledge. Stationary communication entities such as libraries, movie theaters, theaters, concert halls or universities are increasingly being marginalized; we no longer go to the information; the information comes to us. It

stellt: »In banger Erwartung sahen wir dem Jahr 1984 entgegen. ... Aber wir hatten vergessen, daß es neben Orwells düsterer Vision eine zweite gegeben hatte – ein wenig älter, nicht ganz so bekannt, ebenso beklemmend: Aldous Huxleys Schöne Neue Welt. ... Orwell warnte vor der Unterdrückung durch eine äußere Macht. In Huxleys Vision dagegen bedarf es keines Großen Bruders, um den Menschen ihre Autonomie, ihre Einsichten und ihre Geschichte zu rauben. Er rechnete mit der Möglichkeit, daß die Menschen anfangen, ihre Unterdrückung zu lieben und die Technologien anzubeten, die ihre Denkfähigkeit zunichte machen. Orwell fürchtete diejenigen, die Bücher verbieten. Huxley befürchtete, daß es eines Tages keinen Grund mehr geben könnte, Bücher zu verbieten, weil keiner mehr da ist, der Bücher lesen will. Orwell fürchtete jene, die uns Informationen vorenthalten. Huxley fürchtete jene, die uns mit Informationen so sehr überhäufen, daß wir uns vor ihnen nur in Passivität und Selbstbespiegelung retten können. Orwell befürchtete, Wahrheit könnte vor uns verheimlicht werden. Huxley befürchtete, die Wahrheit könnte in einem Meer von Belanglosigkeit untergehen. ... In 1984 werden die Menschen kontrolliert, indem man ihnen Schmerz zufügt. In Schöne Neue Welt werden sie dadurch kontrolliert, daß man ihnen Vergnügen zufügt. Kurz, Orwell befürchtete, das, was uns verhasst sei, werde uns zugrunde richten. Huxley befürchtete, das, was wir lieben, werde uns zugrunde richten.«[8]

Neil Postmans Buch handelt von der Möglichkeit, daß Huxley und nicht Orwell recht hatte. Auch wenn Postmans Kulturpessimismus heute etwas antiquiert wirkt, so sollten wir doch zumindest die darin aufscheinende Dialektik des technischen Fortschritts bedenken. Gerade die Geräte der i-Familie von Apple sind mit ihrer Verführungspotenz ein glänzendes Beispiel für diese Dialektik: Wir können, wenn wir wollen, uns wehren. Aber wir wollen nicht. Das wäre ja zu schön, um wahr ..., um Apple zu sein.

has become freely floating, mobile, ubiquitous and available anytime, anywhere. This creates a new form of intimacy, which, however (as the flipside of the coin) is also increasingly controlled, monitored and used for personal interests by »data octopuses« such as Google, Microsoft and, increasingly, Apple. This surveillance is more a massaging of the consumer in the sense of Aldous Huxley than brutal oppression in the sense of George Orwell.

In 1986 already, in the light of the abundance of entertainment offers, Neil Postman set this judgment against the background of these two utopian scenes in his book *Amusing Ourselves To Death* with the striking subtitle *Public Discourse in the Age of Show Business:* »We anxiously awaited the year *1984*. … But we had forgotten that there was a second vision alongside Orwell's somber one – a bit older, not quite as famous, but just as nightmarish: Aldous Huxley's *Brave New World*. … Orwell warned us of the oppression by an outward force. In Huxley's vision, however, there is no need for a Big Brother to rob people's autonomy, insight, and history. He considered the possibility that people would start to love their oppression and to adore the technology that destroys their ability to think. Orwell was afraid of those who would ban books. Huxley feared that one day there would be no reason to ban books because nobody would want to read them anyway. Orwell was afraid of those who keep information from us. Huxley feared the ones who flood us with information to an extent that we can only save ourselves from them through passiveness and self-reflection. Orwell was afraid that truth could be hidden from us. Huxley feared that truth could drown in an ocean of triviality. … In *1984* people are controlled by causing them pain. In *Brave New World* they are controlled by causing them joy. In short: Orwell was afraid that what we hate would destroy us, while Huxley feared that what we love would destroy us.«[8]

Neil Postman's book is about the possibility that Huxley, not Orwell, was right. Although Postman's cultural pessimism seems a bit outdated today, we should at least consider the dialectics of technological progress that appears in it. Especially the devices of Apple's i-family with their potency to seduce are a shining example of this dialectics: we can, if we so choose, defend ourselves. But we don't want to. That would be too good to be true …, to be Apple.

11. Applemania, Frankfurt, 2010.
12. Beatlemania, London, 1965.

Externe Voraussetzungen: Zur Entwicklung digitaler »consumer electronics«

Die Miniaturisierung elektronischer Bauteile ermöglichte ab dem Ende der 1970er Jahre transportable Audio- und Videospielgeräte, von denen der Walkman und der Gameboy am bekanntesten wurden. Die Walkmans begründeten die Welle der mobilen elektronischen Kleingeräte.[9] Ab etwa 1980 erreichte ihr Absatz bei der primären Zielgruppe der 15- bis 30jährigen erstmals nennenswerte Stückzahlen. Sie richteten sich an eine mobilitätssüchtige Klientel von Joggern und Aerobicfans, Bergwanderern und Fahrradfahrern. Von Generation zu Generation schrumpften die Geräte immer mehr: von doppelter Zigarettenschachtelgröße auf Kassettengröße. In den 1980er Jahren wurden knapp 50 Millionen Walkmans verkauft, in den 1990er Jahren eine ähnliche Anzahl von Gameboys, auf denen man ortsunabhängig Geschicklichkeitsspiele spielen kann. Beide Geräte wurden zum Symbol einer ganzen Generation: »klein, handlich, sportlich und immer in Bewegung.«

Bald schon waren diese Taschengeräte erschütterungsresistent und konnten damit die bis dahin nicht eben seltenen »Aussetzer« kompensieren. Neben Sony, dem Erfinder der Walkmans, brachten bald Firmen wie Panasonic und Philips, Sharp und Samsung, Aiwa und Kenwood Konkurrenzprodukte auf den Markt; eine Diversifizierung und Filialisierung, die 25 Jahre später auch die Mobiltelephone und speziell die Produkte von Apple charakterisiert. Ab 1982 kamen die Discmans dazu, nachdem die von Sony und Philips entwickelte CD-Technik marktreif geworden war.

Ab 1995 folgten die mobilen Minidisc-Recorder, bei denen die CD nur noch einen Durchmesser von 7 cm hatte. Heute sind diese Kassetten- und CD-Geräte seit der Erfindung und Verbreitung der MP3-Player allenfalls noch Flohmarktware. Die Musik »entortete« sich mit diesen Gerätegattungen und korrespondiert damit Paul Virilios zeitgleichen telematischen Visionen. Auch Vilém Flusser hat dies betont und spricht von neonomadischen Existenzen, deren Fokus sich von Besitz auf Zugangschancen umstellt. So emanzipierte sich die Musik ein weiteres Mal von realen Tonträgern ebenso wie von den Anbietern dieser Tonträger, den Plattenfirmen, durch das elektronische Herunterladen von MP3-Dateien. MP3 ist ein digitaler Kompressionsstandard, der Sound-Dateien so weit verdichtet, daß sie sich problemlos über das Internet verschicken lassen.[10] Anfang 1998

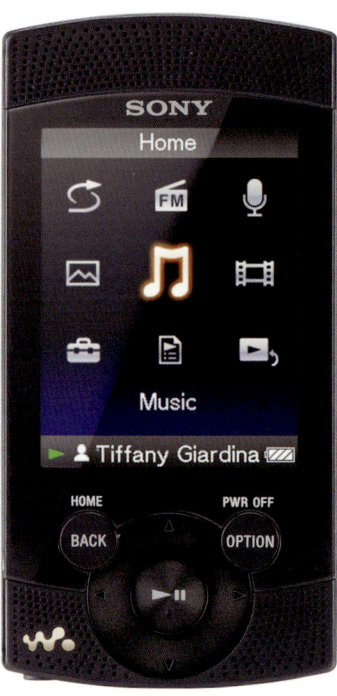

External preconditions: On the development of digital consumer electronics

By the end of the 1970s the miniaturization of electronic components enabled the production of portable audio and video devices of which the Walkman and Gameboy were the most famous examples. The Walkmans started the tsunami of small mobile electronic devices.[9] Around 1980 their sales in the primary target group of 15 to 30 year-olds for the first time achieved numbers worthy of mention. They targeted joggers and aerobics fans, mountain hikers and bicyclists with an addiction to mobility. The devices got smaller and smaller with each generation, from the size of two packs of cigarettes to the size of an audio cassette. Almost 50 million Walkmans were sold in the 1980s, and during the 1990s a similar number of Gameboys, which allowed people to play games of skill independent from a location, were sold. Both devices became defining symbols of an entire generation: »small, handy, sporty, and always on the move.« Soon, these pocket devices became shock-resistant which eliminated the rather frequent »skips«. Alongside Sony, the inventor of the Walkman, companies such as Panasonic, Philips, Sharp, Samsung, Aiwa and Kenwood would soon begin producing competing products; a diversification and filiation, which also characterizes the mobile phones and especially Apple's products 25 years later. Starting in 1982 the Discmans were added after the CD technology, which Sony and Philips jointly developed, had achieved market readiness. The mobile minidisc recorders for CDs with a diameter of only 7 cm followed in 1995. Today, these cassette and CD devices have at best become flea market items since the arrival of MP3 players. With these device genres, music »delocalized« itself and thus corresponds with Paul Virilio's simultaneous telematic visions. Vilém Flusser also emphasized this and speaks of neo-nomadic existences where the focus shifts from possession to access opportunities. Music thus emancipated itself one more time from real storage media as well as from the suppliers of these media, the recording companies, through the electronic download of MP3 files. MP3 is a digital compression standard that reduces the size of audio files without adversely affecting audio quality allowing the files to be transfered via the Internet without any problems.[10]

In early 1998 the Elger Labs presented the first MP3 player MPMAN F10 with 32 MB of storage capacity. A few months later, in September 1998, the

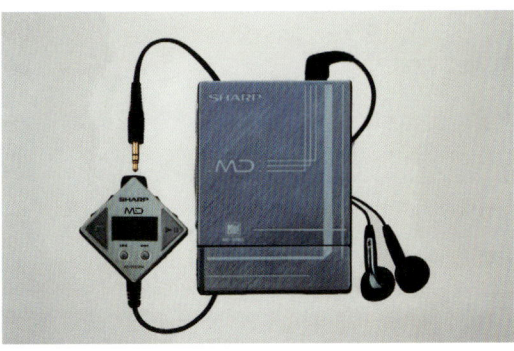
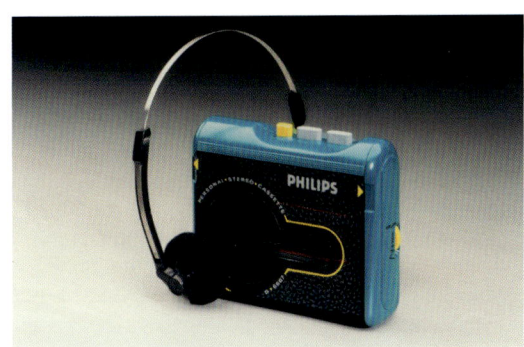

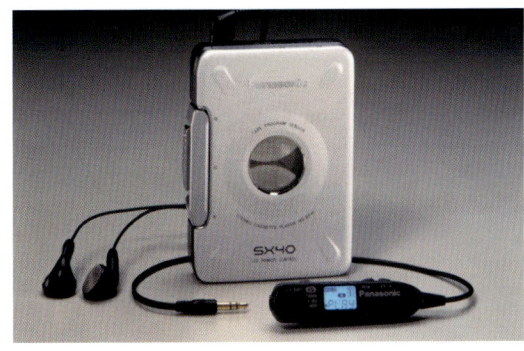

13. Walkman, Sony, 1979.
14. Walkman S540, Sony, 2009.
15. Walkman DD I, Sony, 1989.
16. Radio cassette player AQ6860, Philips, 1980.
17. Cassette player, Sharp, 1989.
18. Cassette player D6607, Philips, 1987.
19. Cassette player SX 85, Panasonic, 1989.
20. Cassette player SX 40, Panasonic, 1989.

 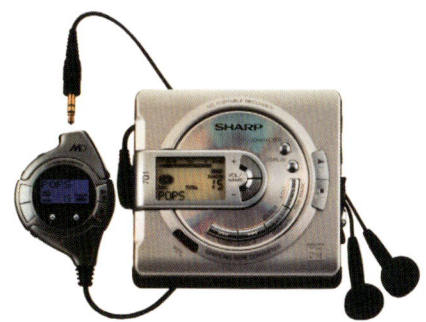 

stellten die Elger Labs den ersten MP3-Player MPMAN F10 mit 32 MB Speicherkapazität vor. Schon wenige Monate später, im September 1998, folgte der schwarze Rio PMP 300 mit LCD-Bildschirm und einem runden Tastenfeld der Firma Diamond Multimedia Systems Inc. für ca. 400 DM. Das runde Bedienungsfeld dieses muskulär wirkenden Audio-Players nimmt das Click Wheel-Rad des iPod nano um sieben Jahre vorweg. Vor der Markteinführung wollte ein kalifornisches Bezirksgericht den Verkauf des Rio verbieten. Die RIAA (Recording Industry Association of America) sah die Rechte des »Audio Home Recording Act« verletzt. Aber die Plattenindustrie unterlag. Nach dem Ende des Rechtsstreits verkaufte Diamond Inc. immerhin 200 000 Geräte.[11] Der Rio konnte zwischen acht und zwölf Stunden Musik speichern, also 100 bis 150 Songs. Heute, zehn Jahre später, beträgt die Speicherkapazität von MP3-Playern locker 30 000 Songs, also circa 2000 Stunden, was etwa 80 Tagen entspricht.

Natürlich bedeutet die Möglichkeit des Herunterladens aus dem Internet die größte Umwälzung auf dem Musikmarkt seit der Erfindung der CD. Genau deshalb hatte ja, dies wissend, die Plattenindustrie den Rio-Prozeß angestrengt. Branchenschätzungen zufolge waren 2001 rund 90 % aller privaten Überspielungen von im Netz eingestellten Songs illegal.

Vorher schon hatte seit Mitte der 1990er Jahre vor allem in der juvenalen Kommunikation die Miniaturisierung neue Geräte ermöglicht, wie etwa die Beeper und Pager. Es handelte sich dabei um mobile kleine Geräte, die auf der Basis von Funk-Rufdiensten funktionieren und flächenbegrenzt Nachrichten senden oder weitergeben können. In Deutschland verkauften drei Unternehmen diese meist scheckkarten- oder tennisballgroßen Geräte: die Deutsche Telekom Skyper und Scall, Motorola das TelMi und die Firma Miniruf das Quix. Schon die Namen sind ein Hinweis darauf, daß diese damals zwischen 100 und 150 DM teuren Funk-Ei-

erchen als Propädeutik für professionelle Mobiltelephone und den damit verbundenen Kommunikationsgewohnheiten verstanden werden können. In dem Maße jedoch, wie die Handys, auch durch einen drastischen Preisverfall, bei Jugendlichen immer mehr Verbreitung fanden, verschwanden die Beeper und Pager vom Markt. Am Ende der 1990er Jahre waren alle diese Kleingeräte praktisch vom Markt verschwunden: Schon zur Jahreswende 2001 hatten in Deutschland ca. 80 % einer Schulklasse mit Vierzehnjährigen ein Mobiltelephon.

Neuerdings gibt es für die Pager zaghafte Wiederbelebungsversuche. So erhalten seit Sommer 2010 Patienten der Unfallklinik Frankfurt am Main kostenlos einen sog. »Patienten-Pager«. Der Patient kann sich damit im Umkreis eines Kilometers um das Krankenhaus bewegen und wird dann per Funk zur Behandlung bestellt. Dieser Pager ist kostenlos und anmeldefrei.

Im Zusammenhang dieser Geräte, vor allem auch der Personal Computer selbst, war die Erfindung der Displays von geradezu epochaler Bedeutung. Auch hier war Apple der Vorreiter. Der erste Macinthosh-Rechner, 1984 vorgestellt, unterschied sich von allen vorherigen Computern vor allem durch seine graphische Benutzeroberfläche. Über diese Oberfläche mit »Ordnern« und »Papierkorb« auf dem Bildschirm konnte man auf dem Computer ebenso einfach Ordnung schaffen wie auf dem wirklichen Schreibtisch.

Peter Sloterdijk hat über die medienhistorischen, ja medienlogistischen Folgerungen der Display-Technik nachgedacht. So wie die Gutenberg-Technik eine bürgerliche Öffentlichkeit geschaffen habe, so erzeuge die neue Display-Technik eine nachbürgerliche Öffentlichkeit, »die von der Formatierbarkeit, Vielfachbespielbarkeit und Individualisierbarkeit der Monitorflächen unterstützt wird. … Der klassische Leser nimmt Texte auf (Rezeption); der aktuelle Leser fängt Texte ab (Interzeption).

21. CD-Player MTV 77, Aiwa, 1989.
22. MiniDisc-Player MD – ST 60, Sharp, 1990.
23. MPaxx, Grundig, 1999.
24. DAT-Player, Sony 1990.
25. 1st MP3-Player Rio 300, Diamond Multimedia, 1998.
26. Game Boy Micro, Nintendo, 2005.
27. Game Boy, Nintendo, 1990.
28. Game Boy Advance, Nintendo, 2001.
29. Nintendo DS, Nintendo, 2005.
30. Gameboy Printer, Nintendo, 1988.

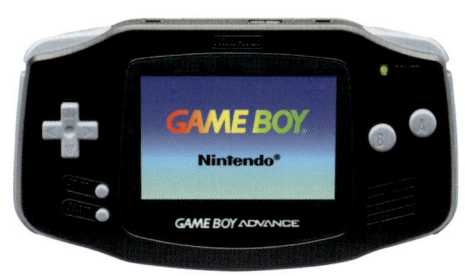

black Rio PMP 300 with LCD screen and a round keypad from Diamond Multimedia Systems Inc. for approximately 400 Deutsche Marks followed suit. The round keypad of this muscular looking audio player anticipated the click wheel of the iPod nano seven years before it appeared on the scene. Before the market launch a Californian district court tried to prohibit the sale of the Rio. The RIAA (Recording Industry Association of America) believed that it violated the »Audio Recording Act«. But the recording industry lost. After the legal dispute was settled, Diamond sold 200 000 of the devices.[11] The Rio was able to store between eight and twelve hours of music, equaling 100 to 150 songs. Today, ten years later, the storage capacity of MP3 players easily equals 30 000 songs or 2000 hours, which is the equivalent of approximately 80 days.

Of course, the possibility to download music from the Internet represents the biggest transformation on the music market since the invention of the CD. It was precisely for this reason that the recording industry filed the Rio lawsuit. According to industry estimates, in 2001 almost 90 % of all songs downloaded from the Internet were illegal.

Before, since the mid-1990s, miniaturization had enabled new devices especially in youth communications such as the beepers and pagers. They were small mobile devices that utilized radio frequencies and call services and could send or forward messages within a restricted area. In Germany, three companies sold these credit card or tennis ball sized devices: Deutsche Telekom sold Skyper and Scall, Motorola the TelMi, and Miniruf the Quix. The names already hinted at the fact that these small radio »eggs«, which cost between 100 and 150 DM at the time, could already be viewed as propaedeutics for professional mobile phones and the always-connected communication habits. However, the beepers and pagers disappeared from the market at the same rate that mobile phones – also due to a drastic price collapse –

spread among the younger population. By the end of the 1990s all of these small devices had practically disappeared from the market: at the turn of the year 2001 approximately 80 % of all fourteen-year-olds in Germany owned a mobile phone.

Recently, there have been tentative attempts to revive the pager. Starting in the summer 2010 patients of the Casualty Hospital Frankfurt am Main will receive a so-called »patient's pager« free of charge. Patients will be able to wander up to one kilometer from the hospital and will be informed via the pager when it is time for their treatment. This pager is free of charge and does not need to be registered.

In the context of these devices, above all the personal computer itself, the invention of the displays was of almost epochal importance. Here, too, Apple was the pioneer. The first Macintosh computer, presented in 1984, differed from all previous computers mainly due to its graphic user interface. This interface with »folders« and a »trash bin« on the monitor made it as simple to clean up and organize one's computer as an actual desk.

Peter Sloterdijk has pondered the media-historical, even media-logistical conclusions of display technology. Just as Gutenberg's moveable type created a civil public, the new display technology has created a post-civil public »that is supported by the ability to format, the multiple recordability and individualization of the monitor surfaces. … The classic reader receives texts (reception); the current reader intercepts texts (interception). This corresponds with the change of accentuation, which is characteristic for contemporary culture as a whole, from the interest in storage to the interest in processing. … Due to these stipulations it is important for future media to participate in the display revolution. This can be recognized in current mobile phone design, for example, which is inconceivable without micro-monitors. Because of the monitor – no matter how miniaturized – the

Dies entspricht dem für die Gegenwartskultur im ganzen charakteristischen Akzentwechsel vom Interesse an Speicherung zum Interesse an Verarbeitung. ... Aufgrund dieser Vorgaben ist es für künftige Medien von Bedeutung, an der Display-Revolution teilzunehmen. Man kann das exemplarisch am aktuellen Mobilephone-Design erkennen, das ohne Mikromonitor nicht mehr denkbar ist. Durch den Monitor, und wäre er noch so miniaturisiert, gewinnt das entwickelte Mobiltelephon Anschluß an den Datenraum, somit an das, was unter aktuellen medialen Bedingungen als das Öffentliche gelten kann.«[12] Vor zehn Jahren geschrieben, liest sich diese Überlegung wie eine geradezu prophetische Vorwegnahme der Gerätecharakteristik von iPod, iPhone und iPad.

Überhaupt werden mit zunehmend ausgereifter Technik die Geräte immer kleiner, nicht nur in zwei, sondern in drei Dimensionen. Man kann sagen, daß im Progreß der Moderne die Volumen förmlich implodieren. So geht etwa die Entwicklung von der großen Plattenkamera zur miniaturisierten Digitalkamera, vom stationären Schnurtelephon mit »Knochenhörer« zum kleinen und flachen Mobiltelephon, vom großvolumigen Röhrenfernseher zum Plasma-Flachbildschirm, vom kühlschrankgroßen Elektroherd zur Mikrowelle, von Glühbirnenleuchten mit Gehäusen zu winzigen Minibirnchen auf gespannten Drahtseilen, vom raumgroßen Straßenkreuzer zum Kleinwagen. Dabei verkennen wir nicht die sozialen und ökologischen Implikationen und Folgerungen solcher technischen Entwicklungen, aber der technische Fortschritt ermöglicht sie nicht nur, sondern verändert über sie auch unser generelles Verständnis von instrumenteller Vernunft. Avancierte Techniken brauchen nicht mehr »die Muskeln spielen zu lassen« oder gar etwas »beweisen«, sondern kultivieren eine Ästhetik der Beiläufigkeit, die das Exzeptionelle als gelassene Selbstsicherheit positioniert. Demgegenüber sind die aufgemotzten SUVs (Sport Utility Vehicles) ebenso eine parvenuehafte Verirrung wie die kantige Formensprache der hochpreisigen Vertu-Mobiltelephone, die zwischen Art Deco und Stealth-Bomber changiert.

Insofern sind diese »Scheich«- und »Oligarchen«-Handys die SUVs der Mobiltelephone. Dem widerspricht nicht das Mirakulöse der Touchscreen-Displays. Denn bei ihnen müssen ja alle denkbaren Inhalte, also »Gestaltungen« erst aktiviert werden, ehe ihre Attraktivität sichtbar wird. Vorher sind die Displays leere, schwarze Flächen, in einem direkten Sinn »unbeschriebene Blätter«. Im übrigen wächst damit dem designtheoretischen Begriff der Black Box eine erweiterte Bedeutung zu: Hinter den Displays verbergen sich nicht mehr nur eine begrenzte Anzahl von Handhabungsoptionen wie noch bei manchen unterhaltungselektronischen Geräten der 1970er und 1980er Jahre, sondern buchstäblich hunderte Millionen von Handlungs-, Unterhaltungs- und Informationsoptionen. Der Datenraum hat keine räumlichen Dimensionen, ihm fehlt gewissermaßen die aristotelische Einheit von Raum, Zeit und Ort.

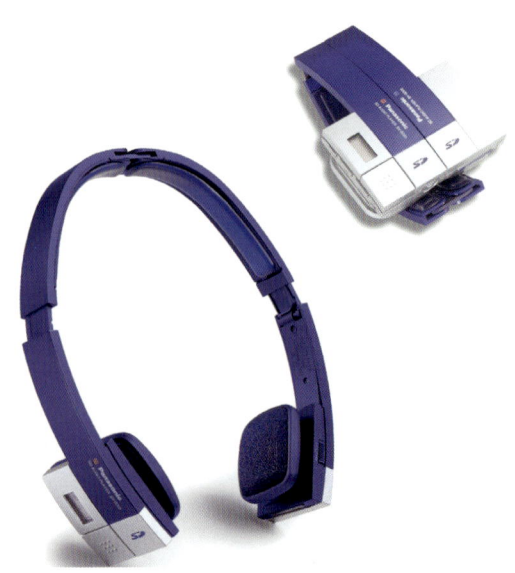

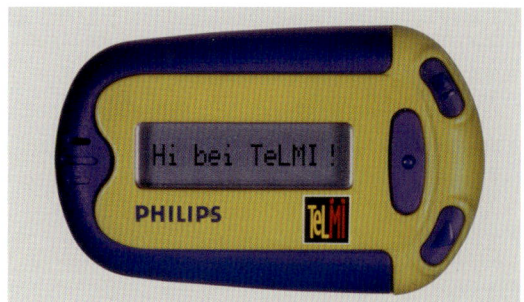

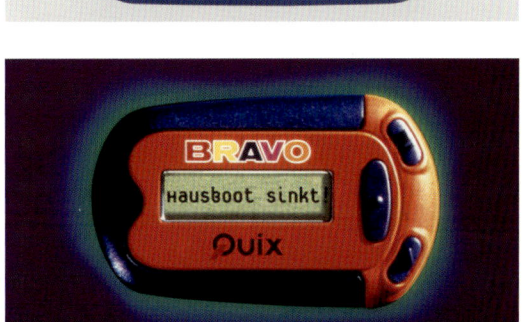

31. Discman with speakers, Sony, 1988.
32. Sony UKW/MV receiver with mobile radio unit, Sony inhouse design, 1981.
33. Foldable audio player as earphones SV-SD 05 E, Panasonic, 2001.
34. Patient pager, Apronti, 2010.
35. TelMi, Motorola, 1996.
36. TelMi, Philips, 1997.
37. Quix, Bravo, 1997.
38. Quix, MTV, 1997.

developed mobile phone gains its connection with the data realm, and hence to what can be considered the public under current media conditions.«[12] Written ten years ago, this consideration almost reads like a prophetic anticipation of the iPod, iPhone, and iPad.

At any rate, with increasingly sophisticated technology the devices become smaller and smaller, not only in two dimensions but in three. One can say that the volumes literally implode in the evolution of modernism. Technology evolves, for example, from large format plate cameras to miniaturized digital cameras, from stationary wired phones with »bone-shaped« telephone handsets to small, flat mobile phones, from the large, bulky CRT television sets to flat screen plasma and LED TVs, from refrigerator-size electric ranges to microwave ovens, from incandescent light bulbs to tiny LEDs inside translucent cable and multi-mirror bulbs on tight wires, from room-size highway cruisers to subcompact cars. We do not misjudge the social and ecological implications and conclusions of such technological developments; however, the technological progress not only enables them but, beyond them, also changes our general understanding of instrumental rationality. Advanced technologies no longer need to »flex their muscles« or even »prove« something but cultivate an aesthetic of arbitrariness, which positions the exceptional as relaxed self-assurance. Compared with this the pimped out SUVs (sport utility vehicles) are rather parvenu-like aberrances such as the edgy formal language of the expensive Vertu mobile phones, which alternates between art deco and stealth bomber. In this respect, these »sheikh« and »oligarchy« phones are the SUVs of mobile phones. The miracle of the touchscreen display does not contradict this. In this case, all conceivable content, i.e., »designs«, first has to be activated before the attractiveness becomes visible. The displays are otherwise void, black surfaces, in a direct sense »blank slates« or »dark horses.« Besides, the theoretical design term black box is given an expanded meaning: no longer are just a limited number of handling options hidden behind the displays, as was the case with some electronic entertainment devices of the 1970s and 1980s, but literally hundreds of millions of handling, entertainment, and information options. The data realm has no spatial dimensions, in a sense it lacks the Aristotelian unity of space, time, and place.

Interne Voraussetzungen: Das Entstehen der Applemania

Sie gehört zu den am häufigsten erzählten Mythen der jüngeren Industriegeschichte: die Erzählung vom Entstehen und der Gründung des Unternehmens Apple Computer Inc., welches seit dem verstärkten Engagement der Firma im Unterhaltungssektor seit 2007 nur noch Apple Inc. heißt. SO erscheint es ausreichend, diese Geschichte hier nochmals nur in aller Kürze zu rekapitulieren.

1976 tüftelten zwei junge Informatik-Studenten, Steve Jobs und Steve Wozniak, in einer Garage in Los Altos im Silicon-Valley in Kalifornien mit einem Startkapital von 1750 US-Dollar den ersten Rechner Apple I aus, der für knapp 670 US-Dollar mit dem Slogan »Byte into an Apple« einige hundert Mal verkauft wurde. Dies war eine naturholzverkleidete Kiste, die wie eine alte Reiseschreibmaschine vom Flohmarkt aussah. Heute steht sie im Smithonian Institute in Washington. Jobs und Wozniak boten ihre Erfindung dem damaligen Großrechner-Giganten IBM an, der aber ablehnte. Dies war mit Sicherheit einer der teuersten und folgenreichsten Fehlentscheidungen der Wirtschaftsgeschichte der Nachkriegszeit.

Erst 1981 brachte IBM seinen ersten eigenen Personal Computer auf den Markt. Bereits 1977 wurde Apple in eine Gesellschaft umgewandelt,[13] nachdem der Apple II sich knapp zwei Millionen Mal verkauft hatte. Er wurde mit einer Broschüre beworben, deren Titelblatt einen frei gestellten roten Apfel zeigte, über dem zu lesen war: »Simplicity is the ultimate sophistication.«

1979 sah Steve Jobs bei einem Rundgang durch das Palo Alto Research Center (PARC) eine Demonstration der graphischen Benutzeroberfläche Alto, die nach dem 1974 entstandenen Alto-Computer benannt ist, der in diesem Research Center entwickelt worden war, einem Grundlagenforschungslabor des amerikanischen Kopierherstellers Xerox: »Ich war geblendet von der graphischen Bedienschnittstelle. Mir schien es das Beste zu sein, was mir je begegnet war, und innerhalb von zehn Minuten war mir klar, daß eines Tages alle Computer so arbeiten würden.«[14] Ab 1981 hatte dann Xerox zwar graphische Benutzeroberflächen zur Serienreife gebracht, verschlief dann aber die weitere Entwicklung, weil die Firma das kommerzielle Potential dieser Technik fälschlicherweise für marginal hielt. So konnte sich Jobs die Rechte an dem Alto-System, auch PARC genannt, sichern, aus dem dann Apple das erste kommerzielle Betriebssystem entwickelte. 1983 kam damit der Nachfolger des Apple II, genannt LISA (Local Integrated Software Architecture) auf den Markt. Erst durch diese graphischen Benutzeroberflächen wurden die PCs massenmarktfähig, denn nun waren keine speziellen Programmierkenntnisse mehr erforderlich.

1984 entwarf Hartmut Esslinger, der CEO des internationalen Designbüros frog design, den Apple Macintosh mit externer Tastatur und Maus, den seine Benutzer – inzwischen eine verschworene Gemeinde – nur Mac nannten. Das Gehäuse war erstmals in gebrochenem Weiß gestaltet. Diese

Internal preconditions: The emergence of applemania

One of the most often told myths of more recent industrial history is the story of the emergence and the foundation of the company Apple Computer Inc., which has been named Apple Inc. since the company became increasingly engaged in the entertainment sector in 2007. In this regard, it seems sufficient to recapture this story in a nutshell here. In 1976 two young computer science students, Steve Jobs and Steve Wozniak, designed and built the Apple I computer in a garage in Los Altos in California's Silicon Valley with startup capital of US $1,750. A few hundred computers were sold for almost $ 670 each with the slogan »Byte into an Apple«. It was a wooden box, with a keyboard that looked like an old travel typewriter from a flea market. Today, one of them is on display at the Smithsonian Institute in Washington. Jobs and Wozniak offered to sell their invention to the computer giant IBM, but the offer was rejected. Certainly, this was one of the most expensive and consequential mistakes in post-war economic history.

The first personal computer produced by IBM arrived on the market only in 1981. By then, Apple had already sold around 20 million PCs. Apple was registered as a corporation in 1977[13] after the company had sold nearly two million Apple II computers. It was promoted with a brochure whose cover presented a floating red apple headed by the slogan: »Simplicity is the ultimate sophistication.« When taking a tour through the Palo Alto Research Center (PARC) in 1979, Steve Jobs saw a demonstration of the graphic user interface Alto, which is named after the Alto computer developed in 1974 this research center, a basic research lab of the American copying machine manufacturer Xerox: «I was dazzled by the graphic user interface. It was the best I had ever seen and within ten minutes I realized that one day all computers would work like that.«[14] Starting in 1981, Xerox had brought graphic user interfaces to series-production readiness; however, the company failed to keep up with further developments because the company wrongly gauged the commercial potential of this technology as marginal. Jobs was thus able to secure the rights to the Alto system, also called PARC, from which Apple developed the first commercial operating system. In 1983 LISA (Local Integrated Software Architecture), the successor to the Apple II, was launched. PCs were ready for the general market because of these graphic user interfaces only after this development; they no longer required the user to have special programming knowledge.

In 1984, Hartmut Esslinger, CEO of the international design office frog design, designed the Apple Macintosh with an external keyboard and mouse, which its users – meanwhile a devoted community – simply called the Mac. The housing for the first time was designed in off-white. This color, called Snow White in the USA, was not least responsible for the cult status of these PCs. For the market launch, Steve Jobs posed on the cover of a Macworld issue with three of these devices. To date, the color white – leaving aside a few excursions into the lollypop colors of the iMacs, iBooks and eMates – has remained a dominant characteristic of Apple computers, notebooks and accessories such as headphones. This white color gentrified the look of computers, which had until then been settled in the industrial working world and usually came in a rather dull gray or beige. The white made PCs aesthetically compatible for living environments and psychologically emphasized the user-friendly menu navigation.[15] Diverse generations of Apple computers were developed over short intervals. Since 1998 six generations of iMac computers alone have been created.[16] The first iMac, like the iBook designed by Jonathan Ive like the iBook, stood out due to a softened, organoid design in semi-transparent candy colors. It finally shifted the design qualities of computers into the realm of fashionable lifestyle products. In 1996 Apple launched the small, translucent eMate computer, which was conceived for schools. It had an integrated pen for writing on the touch-sensitive display, just like Apple's messagepad Newton, whose operating system the eMate used. Later generations of the iMacs returned to the color white.

39. Hartmut Esslinger: Apple IIC, Apple, 1984.
40. Personal computer, IBM with FlexyDisk 5.25 Zoll, 1981.
41. Gavin Ivester, Tim Parsey, Daniele Delulis, Susanne Pierce, Robert Brunner: PDA Newton, Apple, 1993.
42. frog design and Apple (Thomas Meyerhöffer, John Tang, David Baik): Portable PC eMate 300, Apple, 1996.

Farbe, in den USA Snow White genannt, war nicht zuletzt für den Kultstatus dieser PCs verantwortlich. Zur Markteinführung posierte Steve Jobs mit dreien dieser Geräte auf dem Titelblatt einer Macworld-Ausgabe.

Bis heute ist die Farbe Weiß, von einigen Ausflügen in die Lollipop-Farben der iMacs, iBooks und eMates abgesehen, dominantes Merkmal von Apple-Computern, -Notebooks und Zusatzteilen wie etwa Ohrhörern geblieben. Dieses Weiß verbürgerlichte die Anmutung der bis dahin eher in den industriellen Arbeitswelten beheimateten Computer, die in der Regel in stumpfem Grau daherkamen. Das Weiß machte die PCs ästhetisch wohnraumkompatibel und unterstrich psychologisch die nutzerfreundliche Menüführung.[15]

In kurzen Abständen wurden diverse Generationen von Apple-Computern entwickelt. So gibt es seit 1998 bis heute allein sechs Generationen von iMac-Computern.[16] Der erste iMac, entworfen von Jonathan Ive, fiel ebenso wie das iBook durch eine gesoftete, organoide Gestaltung in semitransparenten Bonbonfarben auf. Damit wurden die Anmutungsqualitäten von Computern endgültig in den Bereich modischer Lifestyle-Produkte gerückt. 1996 hatte Apple den transluzenten, kleinen eMate-Computer lanciert, der für Schulen konzipiert war. Er hatte ebenso einen integrierten Stift zur Display-Beschriftung wie Apples Message Pad Newton, dessen Betriebssystem der eMate verwendet. Spätere iMac-Generationen kehrten wieder zum Weiß zurück.

PCs waren seitdem ähnlich hip wie Armbanduhren von Swatch oder Küchen-Equipment von Alessi. Damit fand eine marketingstrategische Neupositionierung statt, die natürlich von den vorausgegangenen elektronischen Mobilprodukten wie Walk- und Discmans, Gameboys, Beepern und Pagern profitierte, die allesamt zunehmend semitransparent gestaltet waren. Im affektiven Koordinatensystem der PC-Bewertungen zog der Spaßfaktor mit dem Arbeitsfaktor gleich. Dies hatte neben der Gehäusegestaltung natürlich auch mit den Inhalten des Internets zu tun, indem jede Art von Wissen, aber auch jede Art von Entertainment abrufbar ist. Und diese Inhalte sind inzwischen gleichermaßen auf den i-Geräten wie den PCs zu empfangen. Der Synergieeffekt der Software korrespondiert mit einem Synergieeffekt der Hardware. Alle Apple-Geräte sind miteinander kompatibel. Allgemeiner gesagt: Die stationären PCs werden trotz Flachbildschirmen und externen Tastaturen über kurz oder lang von den mobilen Geräten verdrängt sein, die ausnahmslos berührungssensitive Touch-Geräte sein werden. Wir erleben gegenwärtig die Götterdämmerung der Tastaturen und Computermäuse.

Diese »Götterdämmerung« wird durch ein Nachfolgegerät für die Maus in der Größe eines Bierdeckels untermauert. Das im Herbst 2010 auf den Markt gekommene Magic Trackpad von Apple aus Glas und Aluminium kommuniziert über Bluetooth mit den Desktop-Computern der Typen iMac, MacBook Pro und Mac mini. Es übernimmt die Multi-Gesten-Steuerung per Fingerbewegung vom iPhone und iPad. Deshalb wird es auch als Touchpad bezeichnet. Auch dieses Gerät markiert eine Konvergenz. Es ist zwar gewohnheitsstrategisch noch eine Maus, aber ihre Bedienungstechnologie ist anders, gewissermaßen eine »Smartmaus«. Diese prinzipielle Entwicklungsrichtung wird auch nicht durch die sogenannte Magic Mouse von Apple außer Kraft gesetzt, die ebenfalls die Multitouch-Technologie des iPhones nutzt.

Und tatsächlich gibt es bereits einen sogenannten All-in-One-Desktop PC, den Acer Aspire Z5710 mit Multitouch-Technologie. In der Printwerbung wird er beworben mit der Headline: »Vergessen Sie die Maus!« Apple selbst hat ebenfalls bereits für PCs ein Patent für berührungssensitive Steuerung angemeldet.

43. Jonathan Ive: iBook, Apple, 1999.
44. Jonathan Ive: iBook, Apple, 2001.
45. Macintosh PowerBook140, Apple, 1991.
46. Jonathan Ive: Magic Mouse, Apple, 2009.
47. Jonathan Ive: Magic Trackpad, Apple, 2010.
48. Hartmut Esslinger: Macintosh SE, Apple, 1984.
49. Jonathan Ive: iMac G3, Apple, 1998.
50. Jonathan Ive: iBook G4, Apple, 1999.

From then on, PCs were almost as hip as Swatch wristwatches or Alessi kitchen tools. A strategic market repositioning took place, which naturally benefited from the preceding mobile electronic products such as Walkmans and Discmans, Gameboys, beepers and pagers, all of which had an increasingly semi-transparent design. In the affective system of coordinates of PC evaluations, the fun factor closed ranks with the utilitarian factor. Of course, alongside the housing design, this had to do with the content of the Internet, through which any kind of knowledge, but also any kind of entertainment, is accessible.

In the meantime, this content can now be received on both i-devices and PCs. The synergistic effect of the software corresponds with a synergistic effect of the hardware. All Apple devices are compatible with one another. In more general terms: in the short or long term, and despite flat screens and external keyboards, the stationary PCs will be displaced by mobile devices, which will invariably be touch-sensitive devices. We are currently experiencing the »Twilight of the Gods« of keyboards and computer mice.

This »Twilight of the Gods« is underscored by a successor device for the mouse the size of a beer mat. The sleek glass and aluminum Magic Trackpad by Apple was launched in the fall 2010 and communicates with the desktop computers models iMac, Mac Pro and Mac mini via Bluetooth. It adopts the multi-gesture control via finger movements from the iPhone and iPad. Therefore, it is also called a Touchpad. This device, too, marks a convergence. It may still be a mouse in terms of usage or »habit strategy«, but its operating technology is different; it is, in a sense, a »smart mouse«. Apple's Magic Mouse, which also employs the multi-touch technology of the iPhone, is neither able to override this basic direction of the development.

And, in fact, a so-called all-in-one desktop PC – the Acer Aspire Z5710 with multi-touch technology – already exists. In print advertising, it is promoted with the headline »Forget the mouse!« Apple itself has already registered a patent for touch-sensitive controls for PCs.

Der Kosmos der i-Geräte: Der iPod und seine Parallel-, Peripherie- und Klonprodukte

Der iPod kam im Oktober 2001 in die Geschäfte. Mit seiner Vorstellung eroberte Apple Schritt für Schritt den Markt der Digitalmusik. Diesen MP3-Player in der Größe einer flachen Zigarettenschachtel hat Jonathan Ive gemeinsam mit dem Apple-Design-Team entworfen. Als Erfinder des iPod gilt allerdings Tony Fadell, den Apple als Hardware-Entwickler verpflichtet hat.[17] Die Frontfläche hat nur zwei Elemente: ein Farbdisplay und darunter eine Wipptaste in Kreisform, das sog. Click Wheel. Diese Taste ist berührungssensitiv und hat in einem äußeren Kreis, angeordnet wie die vier Himmelsrichtungen, vier Funktionen: oben das Menü, mit dem man ein Objekt sucht, unten die Start-und Pausenfunktion, rechts und links das Vor- und Zurückspulen. Im Zentrum befindet sich als zweiter Kreis die Mitteltaste, mit der man die Playlist, die sich in Interpreten, Alben, einzelne Musiktitel oder Genre aufteilt, auswählt. Ein Antippen der Start-Funktion spielt den Musiktitel ab, ein weiteres Tippen auf die Menü-Position regelt die Lautstärke. Wählt man mit der Menü-Funktion eine Liste und berührt dann zweimal die Start-Funktion, werden alle Musiktitel der Liste nacheinander gespielt. Wählt man im Hauptmenü, welches sich scrollen läßt, die Angabe »Zufällige Titel«, dann werden die Songs in beliebiger Reihenfolge abgespielt. Wie bei den Musik- und Videokassetten gibt es die Funktionen »schneller Vor- bzw. Rücklauf«. Gewohnheitsstrategisch knüpft damit die Bedienungslogistik der iPods an die Kassetten- und Videorekorder an, während das Click Wheel tatsächlich eine Neuentwicklung ist, wesentlich attraktiver und angenehmer als z. B. das Bedienkreuz der Gameboys von Nintendo.

Neben dem iPod classic – inzwischen gibt es sieben Generationen – folgte 2004 der iPod mini, der innerhalb eines Monats über 100 000 Mal vorbestellt wurde. Im Januar 2005 erschien der iPod shuffle, der, kleiner als eine Streichholzschachtel, per Clip an der Hemdentasche zu befestigen ist. Hier erfolgt die Bedienung über das Kabel des Kopfhörers. Im September 2005 kam der iPod nano, ebenfalls mit Display und in sechs Metallicfarben, auf den Markt. Inzwischen gibt es wie bei Prepaid-Handys für das iPod-Musikprogramm iTunes Prepaid-Karten ab 15 Euro. Und mit der neuen Geniusfunktion werden die Wiedergabelisten automatisch generiert. Der iPod nano hat heute 16 GB Speicherkapazität: das reicht für 4000 Songs oder 14 000 Photographien sowie für die Speicherung von sechzehn Stunden Videos. Auch ein FM-Radio ist integriert.

Andy Malloy zitiert in seiner Internet-»Apply Computer Reading List« im Zusammenhang des Buches *The Perfect Thing: How the* iPod *Shuffles Commerce, Culture, and Coolness* von Steven Levy, New York, 2007 eine Einschätzung des Network-Forums BN (= burn notice): »Apple Computer, ein Unternehmen, das – wenn auch nicht unbedingt für seinen dominanten Marktanteil – vor allem für seine schicke, topaktuelle Technologie bekannt ist, brachte ein Produkt mit einem verlockenden Versprechen auf den Markt: Benutzer können eine komplette Musiksammlung in der Tasche mit sich herumtragen. Es wurde iPod genannt. Was dann geschah, überstieg die kühnsten Träume des Unternehmens: Mehr als 50 Millionen Menschen haben die auffälligen weißen Stöpsel dieses Gerätes in ihre Ohren gesteckt, und der iPod ist zur weltweiten Obsession geworden. The Perfect Thing ist vom Design bis zum Marketing und zu seinem verblüffenden Einfluß der definitive Bericht über Apples iPod, dem charakteristischsten Gerät unseres noch jungen Jahrhunderts.«[18]

Vom iPod classic wurden zwischen 2001 und 2008 mehr als 130 Millionen Geräte verkauft. Ab Oktober 2005 war der iPod classic auch videofähig. Ebenso beinhalten diese Kleingeräte Funktionen für Photos, Podcasts und Settings. Spezielle Anwendungsprogramme, sog. Applications, ergänzen die Anwendungsmöglichkeiten. Nochmals intensiviert wurden diese Apps beim iPhone, für

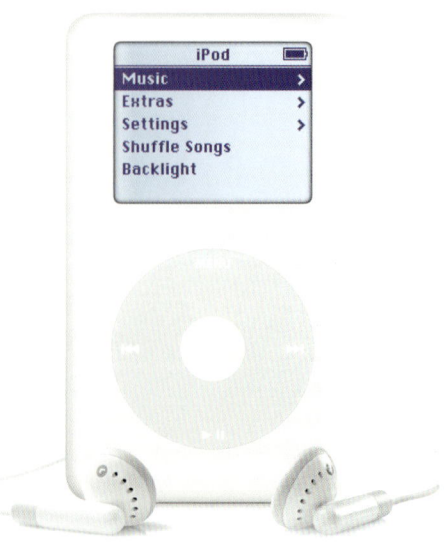

The cosmos of the i-devices: The iPod and its peripheries, filiations, and adaptations

The iPod arrived in stores in October 2001. With its introduction, Apple conquered the digital music market bit by bit. This MP3 player, which was more or less the size of a pack of cigarettes, was designed by Jonathan Ive in cooperation with the Apple design team. However, Tony Fadell, who Apple brought in as a hardware developer, is considered the inventor of the iPod.[17] The front of the device has just two elements: a color display and, below it, a circular navigation element, the so-called click wheel. This click wheel uses both touchscreen-technology and traditional buttons arranged in a circle on the four cardinal points: on the top is the menu button for searching for an object, on the bottom the start and pause function, and on the right and left are the forward and rewind functions. In the center, a second circle, the central button, is for selecting playlists, which are organized by artists, albums, individual tracks or genres. A touch on the start button starts the song; another touch on the menu button regulates the volume. When selecting a list through the menu function and touching the start button twice, all music tracks of the list are successively played. When selecting »random titles« in the main menu, which can be scrolled through, the songs are randomly played. As is the case with music and video cassettes, the functions »fast forward and rewind« are also available. The operational logistics of the iPod is thus modeled after the familiar controls on cassette and video recorders, while the click wheel is in fact an innovation, and it is much more attractive and comfortable than, for example, the cross-shaped directional pad of Ninten-do's Gameboy.

Alongside the iPod classic – now in its seventh generation – the iPod mini followed in 2004; within one month Apple had received 100 000 preorders for this device. In January 2005, Apple launched the wearable iPod shuffle; the device is smaller than a matchbox and can be clipped onto to a shirt pocket. It is operated via a controller on the earphone cord. The iPod nano, which sported a display and came in six metallic colors, was launched in September 2005. As of today, and as is the case with prepaid mobile phones, prepaid iTunes cards are available for purchasing music tracks starting at a price of 15 euro; and playlists can be automatically generated with the new genius function. Today, the iPod nano has a maximum storage capacity of 16 GB, enough for 2000 tracks, 7000 photos or 8 hours of video. An FM radio is integrated into the device as well.

In his »Apply Computer Reading List« Andy Malloy cites an assessment of the network forum BN (= burn notice) in connection with the book *The Perfect Thing: How the iPod Shuffles Commerce, Culture, and Coolness* by Steven Levy, New York 2007: »Apple Computer, a company known for its chic, cutting-edge technology – if not necessarily for its dominant market share – launched a product with an enticing promise: You can carry an entire music collection in your pocket. It was called the iPod. What happened next exceeded the company's wildest dreams. Over 50 million people have inserted the device´s distinctive white buds into their ears, and the iPod has become a global obsession. The Perfect Thing is the definitive account, from design and marketing to startling impact, of Apple's iPod, the signature device of our young century.«[18]

Between 2001 and 2008 more than 130 million iPod classics were sold. Starting in October 2005, the iPod classic was also video compatible. These small devices also include functions for photos, podcasts, and settings. Special applications complete the possibilities of use. The development of these apps was intensified with the release of the iPhone, and currently there are more than 200 000 of these special programs available. Their attractiveness has clearly increased with the almost DIN A4-size iPad. All of the i-devices are about an interaction grammar that is highly user-friendly but

51. Jonathan Ive: iPod classic 1st generation, Apple, 2001.
52. Jonathan Ive: iPod classic 3rd generation, Apple, 2003.
53. Jonathan Ive: iPod classic 4th generation, Apple, 2004.
54. Jonathan Ive: iPod classic 4th generation U2-Edition, Apple, 2004.
55. Jonathan Ive: iPod video 5th generation, Apple, 2005.
56. Jonathan Ive: iPod classic 6th generation, Apple, 2007.

das es über 200 000 solcher Spezialprogramme gibt. Deren Attraktivität ist auf dem fast DIN A4-großen iPad deutlich gesteigert. Bei allen i-Geräten geht es um eine Interaktionsgrammatik, die äußerst benutzerfreundlich, aber durchaus nicht nur einfach intuitiv ist, denn, wie die Kulturwissenschaft weiß: »Jedes Interface ist kulturell bedingt.« Seit der Markteinführung des iPods gibt es bis heute mehr als vierzig Gerätevarianten.

Der iPod ist ebenso wie das iPhone und der iPad ein Speicher für Musik, Bilder, Nachrichten, Filme und Werbung, aber auch für Navigationssysteme, Karten und Branchenverzeichnisse. Dies heißt, daß die i-Geräte viel mehr software- als hardwarebasiert sind. Wirtschaftlich sind sie in erster Linie Marketing-Tools für Platten- und Filmfirmen sowie Printmedien, die hier, oft nicht ganz freiwillig, neue Vertriebswege und Umsätze suchen.

Der iPod mit Bildschirm und Steuerrad ist schnell gattungsbestimmend geworden. Unternehmen wie Archos, Creative, Grundig, Iriver, Philips, Samsung, SanDisk Sansa, Sony, Teac und selbst Tchibo mit seiner Eigenmarke TCM haben mehr oder weniger deutliche Adaptionen kreiert. Der taiwanesische Hersteller Luxpro sorgte Mitte 2005 für Aufsehen, als er den MP3-Player Super Targent vorstellte, der sehr stark an den iPod shuffle von Apple erinnerte. Ein taiwanesisches Gericht beschied allerdings eine Klage Apples abschlägig: Es handele sich mitnichten um ein Plagiat. Ein optischer Vergleich allerdings belegt das Gegenteil. Aber es gibt auch MP3-Player, die dem Flachform-Minimalismus des iPods widerstehen. Dazu gehören ebenso japanische Miniwürfel von Panasonic/Matsushita wie handschmeichlerische Freiformen aus den USA.

Allerdings wäre es falsch, den oft relativ hilflos wirkenden »Geometrismus« vieler iPod-Nachahmungen schon für einen gekonnten Minimalismus zu halten. Denn Hersteller dieser iPod-Mimikry, wie z. B. Grundig mit seinem MP3-Player MPaxx 920, Philips mit seinem Audio-Video-Player SA 3125 oder Tchibo, schaffen es doch immer irgendwie, verschwiemelte gewölbte Teilformen in ihre Geräte zu integrieren, die entweder wie Pastenstränge aus einer Tube oder, wie beim Kippschalter des Gerätes von Tchibo, wie ein abstrahiertes Gänseblümchen aussehen. Mehr noch: Trotz der weltweiten Dominanz der Apple-Geräte gibt es noch extreme gestalterische Peinlichkeiten in Surfbrett- oder Hantel-, in Trapez- oder Eiformen. Dies erinnert an vergleichbare fatale Entwürfe bei Mobiltelephonen: So hatte Anfang der 1990er Jahre z. B. Siemens das »S 45« Handy im Programm, dessen Gestaltung nierenförmig daherkam. Auch Nokia frönte in dieser Zeit solchen Freiformen.

Im übrigen halte ich es für wahrscheinlich – um auf die oft behauptete Analogie zwischen Apple und Braun zu sprechen zu kommen –, daß für die meist jugendlichen Nutzer das asketische Weiß oder Schwarz der iPods eher zweitrangig ist. Wohl kaum zufällig gibt es die iPod nanos in verschiedenen »coolen« Farben, wie die Werbung verspricht: metallic-blau, metallic-lindgrün, metallic-pink, metallic-schwarz bzw. weiß. Auch der iPod classic wird in Softeisfarben angeboten. Dies scheint mir kaum ein Neo-Funktionalismus in der Tradition von Braun zu sein. Viel bestimmender für die Geräte ist die Software iTunes, mit der man sich je nach GB-Kapazität z. B. auf den iPod nano 500 oder 1000 Songs herunterladen kann.

Für beide Geräte wären zudem, zumindest für ihre Kanten und Rückseiten saisonal unterschiedlich ornamentalisierte Oberflächen im Sinne einer »Swatchisierung« denkbar. Für Sonderserien von Laptops ist eine solche Individualisierung, z. B. durch aufgesetzte Swarovski-Kristalle oder andere bunte Ornamentik, längst üblich. Ohnehin gibt es bereits limitierte Sondereditionen des iPod; eine Strategie, die man designtheoretisch »customization« nennt, etwa ab Oktober 2004 die limitierten iPod U2-Editionen mit den gravierten Signaturen der vier Bandmitglieder auf der Rückseite. Im Oktober 2006 stellte Bono von U2 die Sonderedition iPod nano Product red vor. Eine weitere Sonderedition widmete sich 2005 in den USA Harry Potter, inoffizielle Special Editions mit Songs und Videos der Sänger Xavier Naidoo und Mousse T. wurden, ebenfalls 2005, nur über den Apple-Händler Gravis vertrieben.

In welchem Maße die Infiltration des i-Kosmos bei Mitbewerbern gediehen ist, zeigt die Tatsache, daß es rund um die iPods inzwischen »buchstäblich Tausende von Produkten gibt, von externen Lautsprechern über das feine Lederetui und die Moshi-Tasche iPouch bis zur iPod-Socke in grell leuchtenden Farben. … Neuerdings gibt es eine Wurlitzer-Jukebox für den iPod mit Bose-Soundsystem, die läppische 9000 Euro kostet«, wie Michael Spehr in einer Technikbeilage der *Frankfurter Allgemeinen Zeitung* 2007 bemerkte.[19]

Am deutlichsten wird diese Kolonialisierung der Produktportfolios von eigentlich konkurrierenden Mitbewerbern bei den »Docking Stations«. Es ist belegt, daß es innerhalb der Produktgeschichte immer einmal wieder »Schlüsselprodukte« gab, die ihr Hersteller durch kompatible Zusatzgeräte veredelte oder über Lizenzverträge von anderen Herstellern veredeln ließ. Beides ist hier der Fall. So wird auch der zigarettenschachtelgroße iPod classic, welcher zehntausende Songs speichern kann, zusätzlich durch solche vielfach größere »Docking Stations« mit externen Lautsprechern nobilitiert, die einen sonoren Raumklang erzeugen. Das eingeklinkte iPod mutiert solchermaßen – im übrigen tatsächlich wie ein Schlüssel für eine Tür – zu einem üblichen Radiogerät bzw. einer HiFi-Anlage, allerdings mit dem entscheidenden Unterschied, daß alle gespeicherten Songs individuell nacheinander oder über Zufallsstrukturen (sog. Random-Programme) anwählbar sind. Die Abhängigkeit von Senderprogrammen und Sendezeiten entfällt, aber eben auch die Attraktivität und Spannung der Programmsuche. Apple besteht bei solchen Ergänzungsgeräten auf einem Aufkleber, der die Kompatibilität des Gerätes zu Apple-Produkten ausdrückt. Neben einem Piktogramm des iPod steht der Satz »Made for iPod«, der bei manchen Geräten inzwischen durch ein Piktogramm des iPhone und der Sentenz »Works with iPhone« ergänzt wird. Die Unternehmen, die für den iPod und das iPhone »Docking Stations« auf den Markt bringen, müssen an Apple nennenswerte Lizenzbeträge bezahlen. Offenbar motiviert der Besitz eines iPods, iPod touch oder

57. Jonathan Ive: iPod mini 2nd generation, Apple, 2005.
58. Packaging iPod classic 2nd generation, Apple, 2002.
59. Packaging iPod mini 2nd generation, Apple, 2005.
60. Packaging iPod mini, Apple, 2008.
61. Packaging iPod classic 6th generation, 2007.

absolutely not just intuitive because, as cultural studies show, »Every interface is caused by culture.« Since the market introduction of the iPod more than forty versions of the device have been released. The iPod, like the iPhone and iPad, has become a storage device not only for music but also for images, news, movies, and advertising, as well as for navigation systems, maps and business directories. This means that i-devices are based much more on software than on hardware. Economically, they are mainly marketing tools for recording and film companies as well as print media that, often not entirely voluntarily, seek new distribution channels and revenues through the medium.

The iPod with a display and control wheel quickly became formative for its product genre. Companies like Archos, Creative, Grundig, Iriver, Philips, Samsung, SanDisk Sansa, Sony, Teac, and even Tchibo with its own TCM brand have created devices that are clearly more or less adaptations of the iPod. The Taiwanese manufacturer Luxpro caused a sensation in mid-2005 when it presented the MP3 player Super Tangent, which strongly resembled Apple's iPod shuffle. A Taiwanese court, however, rejected a complaint from Apple: it decided that it was not plagiarism. However, a visual comparison confirms that the opposite is true. But there are also MP3 players that resist the flat shape minimalism of the iPod. Among them are Japanese mini-dice by Panasonic/Matsushita and smooth, freeform devices from the USA that conform to the hand like hand charmers.

However, it would be wrong to consider the designs of the often relatively helpless looking »geometrism« of many iPod wannabees to be artful minimalism. Manufacturers of these MP3 players such as Grundig with its MP3 player MPaxx 920, Philips with its audio-video player SA 3125 or Tchibo somehow always succeed in integrating swollen, curved parts that either look like strands of paste from a tube or, as is the case with Tchibo's device, an abstract daisy. Moreover, despite the global dominance of the Apple devices there are even more extreme design embarrassments in surf board, dumbbell, trapezoid or egg-shape forms. This is reminiscent of similarly fatal designs of mobile phones: in the early 1990s, for example, Siemens produced the kidney-shaped »S 45« mobile phone. Nokia, too, indulged in such freeform designs during that time.

Furthermore, I consider it probable – to speak of the oft-claimed analogy between Apple and Braun – that the ascetic white or black of the iPods is of secondary importance to most of the youthful users. It is no coincidence that iPod nanos are available in various »cool colors«, as the advertising promises: metallic blue, metallic lime green, metallic pink, metallic black or white. The iPod classic is also offered in soft ice-cream colors. To me, this hardly seems to be a neo-functionalism in the tradition of Braun. The iTunes software, which allows users to download up to 500 or 1000 songs (depending on the GB capacity) for example to the iPod nano, seems much more decisive.

For both devices, ornamented surfaces in the sense of a »Swatchization« are conceivable at least for their edges and backs. Such individualization has long become customary for special laptop series, for example with the addition of Swarovski crystals or other colorful ornamentation. At any rate, limited special editions of the iPod have already been released, for example the limited iPod U2 editions starting October 2004 with the engraved signatures of the four band members on its back – in design theory, this strategy is called customization. In October 2006 Bono of U2 presented the special edition iPod nano product red. Another special edition was dedicated in 2005 to Harry Potter in the USA, and unofficial special editions with songs and videos by the singers Xavier Naidoo and Mousse T. were also distributed in 2005 exclusively by the Apple dealer Gravis. If ordered via the Internet, individual engravings for the iPod are possible.

The fact that »literally thousands of products« have meanwhile been created for the iPod, »from external speakers, fine leather cases and the Moshi iPouch, to iPod socks in bright, shiny colors« proves how a large number of companies have thrived in the i-cosmos. »Recently, we saw a Wurlitzer jukebox for the iPod with Bose sound system that costs a ridiculous 9000 euro«, Michael Spehr noted in a technology supplement of *Frankfurter Allgemeine Zeitung* in 2007.[19]

This colonization of the product portfolios of usually competing peers becomes most evident in the case of the docking stations. It has been proven that there have always been »key products« in product history, which their manufacturers had ennobled through compatible additional devices or through license agreements by other manufacturers. Here, both apply. The cigarette pack sized iPod classic, which can store tens of thousands of songs, is also ennobled with such usually larger docking stations with external speakers that create sonorous stereophonic sound. In this way, the docked iPod – by the way, it is indeed like a key to a door – mutates into a generic radio device or HiFi system, but with the decisive difference that songs can be selected individually, played in random order or grouped into playlists. The dependence of radio programs or broadcasting hours is eliminated, but also the appeal and excitement of searching for a station that fits one's mood. For such additional devices, Apple insists on a label that expresses the compatibility of the device with Apple products. Next to an iPod pictogram, it displays the sentence »Made for iPod«, which has meanwhile been complemented by a pictogram of the iPhone and the sentence »Works with iPhone« on several devices. The companies that produce docking stations for the iPod and iPhone have to pay considerable license fees to Apple. Obviously, the possession of an iPod, iPod touch or iPhone motivates a considerable number of users to buy such docking stations. In marketing theory, such follow-up purchases are called the »omnibus effect«. Apple is cashing in on this motivational impulse.

Technologically sophisticated docking stations are available from, for example, AEG, Altec, Apple, Audiovox, Avox, Bang & Olufsen, B & W, Bose, Bergmann, Denon, Enmic, Geneva, Finite Elemente, JBL, Lenco, Kenwood, Logitech, Muvid, Onkyo, Parrot, Philips, Pioneer, Sonoro, Sony,

iPhones nicht eben wenige Nutzer, sich eine solche Andock-Station zu kaufen. In der Marketingtheorie bezeichnet man solche Folgekäufe als »Omnibus«-Effekt. Diesen Motivationsschub läßt sich Apple honorieren.

Technisch anspruchsvolle »Docking Stations« gibt es z. B. von AEG, Altec, Apple, Audiovox, Avox, Bang & Olufsen, Bowers & Wilkins (B & W), Bose, Bergmann, Denon, Enmic, Geneva, Finite Elemente, JBL, Lenco, Kenwood, Logitech, Muvid, Onkyo, Parrot, Philips, Pioneer, Sonoro, Sony, T+A, Tivoli und Yamaha, um nur einige zu nennen, die das Minikistchen jeweils zu veritablen Audio-Geräten aufblasen: auch dies ein Indiz dafür, daß der iPod weit mehr Benutzeroberfläche als Gerät ist. Zudem sind diese Andock-Stationen ein nachhaltiges Beispiel für »cross marketing« und »image transfer«. Dieser »image transfer« funktioniert selbst da, wo sich das Unternehmen Apple gewissermaßen selbst kannibalisiert. Denn der jüngste iPod, der iPod touch mit acht Gigabyte ist optisch und funktional eine vollständige Adaption des iPhone mit einem identischen Multitouch-Display, Platz für 1750 Songs bzw. 10 000 Photographien, mit einer Kapazität für 30 Stunden Musik oder zehn Stunden Video. Ebenso wie beim iPhone kann man mit dem iPod touch via WiFi im Web surfen, E-Mails abrufen und auf den App Store sowie den iTunes-Music Store zugreifen. Lediglich die Möglichkeit des Telephonierens entfällt. Ein klassisches Beispiel nicht nur für Konvergenz im Produktdesign, sondern auch ein weiterer Hinweis darauf, daß die Reduktionsästhetik der früheren iPods keineswegs jene dominante Rolle spielt, mit der ihre angebliche Verwandtschaft mit dem Braun-Design begründet wird. Im Umkehrschluß ist dann eine der iPhone-Apps der iPod, nun allerdings nur noch als Software.

Das Design dieser Abspielgeräte bedient alle Stillagen und entspricht keineswegs immer dem gestalterischen Minimalismus der iPods. So erinnern die ovaloiden »Docking Stations« der Hersteller JBL, Onkyo, Denon und Bowers & Wilkins (B & W) an die Kurveneuphorie von Bang & Olufsen, die aufgeständerten Quaderkisten von Geneva, der iCube von Lenco oder der Popcube Mini von Bergmann an frühe Audiogeräte der Firma Braun. Viele Geräte, etwa von Avox und Lenco, Logitech und Tivoli, aber auch von Apple selbst, zeigen offene, unverkleidete Lautsprecherkalotten, die diesen Andockstationen nicht nur ikonographisch einen deutlichen Hightech-Touch geben, sondern sie manchmal auch leicht martialisch wirken lassen. Die gesofteten Geräte von Sonoro dagegen oder die Röhrenform von Altec sind gestalterisch eine Melange aus Braun- und Pop-Design. Das iPod-Lautsprechersystem Zeppelin von B & W erinnert an eine aufgeständerte Honigmelone bzw. an einen Baseball, während das nur ca. 5 cm hohe, aber einen Meter breite Soundboard Hohrizontal des Unternehmens Finite Elemente wie ein Regalbrett mit Audiofunktion daherkommt.

Die von Philippe Starck für das Unternehmen Parrot entworfenen HiFi-Lautsprecher Zikmu, zwei trompetenförmige Standsäulen, ermöglichen durch einen 360-Grad-Rundum-Sound eine ultimative iPod/iPhone-Raumbeschallung. Eine Säule dient als Dock, die Audiosignale werden drahtlos zwischen den beiden Säulen übertragen. Dagegen ist eine Andock-Station von Tchibo formal etwas eigenartig, denn deren Form kann sich irgendwie nicht zwischen überdimensionierter Glühbirne und einem ebensolchen Salzstreuer entscheiden. Ebenso merkwürdig ist die »Docking Station« Edifier Breathe, deren Gestalt zwischen Qualle und Toupet changiert. Eher wie ein Equipment aus dem Film Starwars wirkt der Soulra genannte iPod- und iPhone-Verstärker, der, mit Solarenergie betrieben, wasserabweisend und damit strandkompatibel ist. Eine sehr edle, schicke Abspielstation hat der japanische HiFi-Konfektionär Pioneer im Programm. Die jüngsten dieser Kodo genannten Modelle (japanisch: Herzschlag oder Puls) lassen sogar zwei Apple-Geräte, den iPod und das iPhone, gleichzeitig aufspielen. So kann man sich als Hobby-DJ betätigen und von einem zum anderen Player überblenden. Und selbst in Schreib- und Nachttischleuchten werden Andockstationen integriert, etwa bei der LED-Leuchte Diva

62. Jonathan Ive: iPod shuffle 2nd generation, Apple, 2006.
63. Jonathan Ive: iPod shuffle 4th generation, Apple, 2010.
64. Jonathan Ive: iPod shuffle 4th generation, Apple, 2010.
65. Jonathan Ive: iPod shuffle 1st generation, Apple, 2005.
66. Jonathan Ive: iPod shuffle 3rd generation, Apple, 2009.
67. Flyer MP3-player Super shuffle, Luxpro, 2005.

T+A, Tivoli and Yamaha, to name but a few. They expand the mini box into a veritable audio system: this, too, is an indication of the fact that the iPod is a user interface far more than it is a device. In addition, these docking stations are sustained examples of cross-marketing and image transfer. This image transfer works even where Apple, in a sense, cannibalizes itself. Because the most recent iPod, the iPod touch with 8, 32 or 64 gigabytes of memory, is visually and functionally a complete adaptation of the iPhone, with an identical multi-touch display, space for 1750 songs or 10000 photographs, with capacity for 30 hours of music or 10 hours of video. As is the case with the iPhone, the iPod touch allows online surfing via WiFi, access to e-mail and the app store as well as the iTunes music store. It includes all of the capabilities of an iPhone except the ability to make phone calls. This is not only a classic example of convergence in product design but also another hint that the reduced aesthetics of the earlier iPods does not play a dominant role in explaining their supposed relationship with the Braun design. In reverse conclusion, iTunes is merely an iPhone app, merely installed software that allows the iPhone to function as an iPod.

The design of these playback devices serves all styles and does not at all always comply with the design minimalism of the iPods. The ovoid docking stations by the manufacturers, Onkyo, Denon and Bowers & Wilkins (B & W) remind us of the curvature euphoria of Bang & Olufsen, the floor-mounted cubes by Geneva, the iCube by Lenco, the Popcube Mini by Bergmann or previous audio devices by Braun. Many of them, for example by Avox and Lenco, Logitech and Tivoli, but also Apple itself, show open, uncased speaker calottes that not only iconographically provide these docking stations with a clear high-tech touch but at times make them look slightly martial. The softened devices by Sonoro, however, or the tubular shape of Altec Lansing are in their designs a mélange of Braun and pop design. The iPod speaker system Zeppelin by Bowers & Wilkins reminds us of a floor-mounted honeydew melon or baseball, while the soundboard Hohrizontal by the company Finite Elemente, which is only 5 cm high but one meter wide, looks like a shelf with an audio function.

Philippe Starck designed the ultimate iPod/iPhone stereo sound system with the HiFi wireless speakers Zikmu in the shape of two trumpet-shaped columns for the company Parrot. They offer a 360 degrees surround sound. One column serves as a dock and the audio signals are transmitted wirelessly between the two columns. Contrary to this, a docking station from Tchibo is formally somehow peculiar because its form somehow cannot decide between oversize light bulb and saltshaker. The docking station Edifier Breathe, whose design changes between a jellyfish and toupet, is just as curious. The iPod and iPhone amplifier named Soulra, which runs on solar power and is water resistant and thus compatible for use at the beach, looks like a gadget out of the *Starwars* movies. The Japanese HiFi manufacturer Pioneer offers very sophisticated and chic playback stations. The most recent of these models, named Kodo (Japanese: heartbeat or pulse), even allows two different Apple devices, both the iPod and the iPhone, to be played. Users can act as DJs and fade from one player to the other. Docking stations are even being integrated into desk or nightstand lamps, for example in the LED lamp Diva from Italian manufacturer Rotaliana, or the Silver Seiko conof from Japan. The one-button MP3 player TicToc from Samsung, a box with the size and appearance of a pill dispenser, proves to have gerontological correctness. It only has two functions: when held horizontally, it jumps from one track to the next, and the volume is adjusted by holding it vertically. The industry also does not deprive us of a radio alarm clock with video function via docked i-device, as if the music coming from our alarm clocks didn't already alarm us. But perhaps, after falling asleep in front of the TV, it would be nice to be able to seamlessly continue to watch the action movie upon waking up. Obviously, there are no limits to the free floating design fantasy. We may look forward to nostalgic »people's

des italienischen Herstellers Rotaliana oder bei der Silver Seiko conof aus Japan. Gerontologische »correctness« beweist der Ein-Knopf-MP3-Spieler TicToc von Samsung, eine Box in der Größe und Anmutung eines Tablettenspenders. Er hat überhaupt nur zwei Funktionen: Waagrecht gehalten, springt das Programm von Lied zu Lied, senkrecht wird die Lautstärke geregelt. Auch einen Radiowecker mit Videofunktion via angedocktem i-Gerät enthält uns die Industrie nicht vor. Wobei uns ja bereits die Musikfunktion des Weckers manchmal auf den Wecker ging. Aber es ist doch eine schöne Möglichkeit, direkt nach dem Aufwachen den am Abend vorher wegen Wegdämmerns vor dem TV unterbrochenen Spannungsbogen eines Actionfilmes nahtlos fortzusetzen. Offenbar sind der frei schweifenden Entwurfsphantasie keine Grenzen gesetzt. Auf nostalgische »Volksempfänger« als Andock-Station darf man gespannt sein. Dabei ist die Preisspanne der angebotenen »Docking Stations« enorm: Sie reicht von vierstelligen Eurobeträgen bis zu Gratisangeboten, die etwa der deutsche Verpackungs- und Archivierungshändler Pressel als »Give aways« anbietet. Natürlich sind dies dann ausklappbare Minilautsprechern, denen man ihre Eintagsfliegenhaftigkeit ansieht.

Schon Ende 1981 bot Sony eine Art »Prä-Docking Station« an: einen FM/AM-Empfänger mit einem herausnehmbaren Radioteil, welches dann über Kopfhörer hörbar ist. Ein Walkman konnte zwischen die beiden Lautsprecher eingeklinkt werden. Damals kostete dieses Gerät 450 DM. Einer vergleichbaren Gattungsarchäologie werden wir auch beim iPad begegnen. Gerade solche Geräte, die die Industrie ständig als »atemberaubende Neuigkeiten« anpreist, haben oft eine weit zurückreichende Vorgeschichte.

Zur Konvergenz gehört auch, daß Apple ab November 2006 in Kooperation mit dem Sportschuhhersteller Nike ein spezielles Trainingssystem entwickelte. In die Sohle eines Laufschuhs werden Sensoren integriert, die Streckenlänge, Laufleistung, Geschwindigkeit und Kalorienverbrauch beim Joggen ständig an ein mitgeführtes iPod nano senden. Dies erinnert an jene Armbanduhren von Swatch, die eine integrierte Skipaß-Funktion haben. Dadurch entfallen an den Pistenzugängen die Wartezeiten, denn die Nutzungsberechtigung wird beim Vorbeigehen »eingelesen«.

Zum Kultstatus des iPod tragen nicht zuletzt die weißen Kopfhörer bei, denn sie sind oft das einzige, was man von dem Gerät sieht, welches ja in aller Regel in der Kleidung verstaut wird. Sie sind ein Markenzeichen und Wiedererkennungssymbol des iPod geworden. In der Produktbeschreibung des iPod shuffle heißt es: »Die weißen Ohrhörer zeigen, daß Sie Ihre Musik mit Stil genießen.«

Zur marketingstrategischen Produktdiversifikation gehören zudem Dutzende diverser iPod-Hüllen und -Armbänder aus Kunststoff, Leder, Filz oder Silikon. Apple selbst bietet sog. iPod socks an, geradezu bieder wirkende Strickhüllen, die dem Hightech-Produkt heimelige Gemütlichkeit verleihen. Gestrickt, gewirkt, gehäkelt, ummantelt, geschützt, verborgen: Es gibt buchstäblich kein Material, in das die i-Produkte nicht eingewickelt werden. Dies ist in mehrfacher Hinsicht aufschlußreich: Offenbar halten die Nutzer die Geräte für sensibler als der Hersteller und so bieten Dutzende von Lieferanten entsprechende Überzieher an. Zweitens fungieren solche Hüllen im Sinne einer customization als Indikatoren einer Individualisierung. Wer denkt bei den von Apple selbst angebotenen iPod socks nicht an den Sparstrumpf der Oma? Drittens nobilitieren die Hüllen die Produkte im Sinne eines schützenswerten Gutes. Es gibt sogar umfangreiche Nutzer-Kits mit mehreren Teilen. So bietet die Jeansfirma Levi's ein Arrangement mit Namen Redwire an, bei dem der iPod in einer Tasche am Oberschenkel steckt und mit

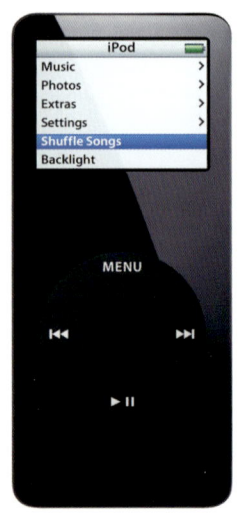 

receivers« as docking stations. The price range of these docking stations is enormous: from more than 10 000 euro to free offers that are used as giveaways, for example by the german packaging and archiving dealer Pressel. Naturally, they are fold-out mini speakers that already look like flashes in the pan.

At the end of 1981, Sony already offered a type of »pre-docking station: »it was a high-performance FM/AM receiver with a removable radio element, which was then operated via headphones. A Walkman could be plugged in between the two speakers. At the time, the device cost 450 DM. We will again encounter a similar genre archaeology in the case of the iPad. Especially those devices that the industry repeatedly advertised as »breathtaking innovations« often have a prehistory reaching far back into the past.

Also part of convergence is the fact that Apple in cooperation with the sports shoe manufacturer Nike developed a special training system back in November of 2006. Sensors integrated into the sole of a running shoe continually send distance, performance, speed, and calorie data to an iPod nano. This reminds us of those wristwatches from Swatch that also have an integrated ski pass function. It reduces a skier's waiting time at the ski-lift because the pass is automatically »read« when the skier passes a sensor.

The white headphones of the iPod also contribute to its cult status because they are often the only part of the device that is visible; the device is usually stowed away in a pocket or pouch. They have become a trademark and recognition symbol of the iPod. The product description of the iPod shuffle states: »The white headphones show that you enjoy your music with style.«

The marketing strategy of product diversification also includes dozens of iPod shells and armbands made of plastic, leather, felt or silicone. Apple itself offers iPod socks, knitted shells that look almost conservative and provide the high-tech product with homey coziness. Knitted, crocheted, encased, protected, and hidden: there is literally no material in which i-products cannot be wrapped. This is revealing in several ways: obviously, the users consider the devices more sensitive than the manufacturer, and hence dozens of suppliers offer the appropriate protective overcoats. Second, such shells function as indicators of individualization in the sense of customization. Who would not think of grandma's stocking in which she kept her savings when looking at the iPod socks offered by Apple? Third, the shells ennoble the product in the sense that it is a commodity that is worth protecting. There are even comprehensive user kits with several parts. The jeans company Levi's offers a line of jeans called Redwire, with an iPod dock built into a side pocket, and a mini joy stick

68. Jonathan Ive: iPod nano 4th generation, Apple, 2008.
69. Jonathan Ive: iPod nano 3rd generation, Apple, 2007.
70. Jonathan Ive: iPod nano 1st generation, Apple, 2005.
71. Jonathan Ive: iPod nano 6th generation, Apple, 2010.
72. Jonathan Ive: iPod nano 2nd generation, Apple, 2006.

einem flachen roten Kabel mit einem isolierten Click Wheel auf der Kleingeldtasche verbunden wird. Die Kopfhörer stecken in einer Gürtelbuchse.

Aber während die textilen Hüllen die i-Geräte gewissermaßen intern vor Beschädigungen schützen, gibt es auch externe textile »Geräte-Erweiterungen« für die i-Familie. So produziert das österreichische Unternehmen Urban Tool modische Taschen, Holster und T-Shirts aus Synthetics für die i-Geräte und andere Gadgets. Der grooveRider z. B. ist ein funktionales Sportshirt mit textiler Interfaceschnittstelle. Von da aus kann ein iPod direkt angesteuert und seine Funktionen können geregelt werden. In einem Kabelkanal werden die Kopfhörer bis an die Schulter geführt. Die beiden Kollektionen von Urban Tool mit etwa vierzig Teilen firmieren unter den Namen iWearables und iPortables. Nun ist es ein Unterschied, elektronische Funktionsteile in Stoffe oder synthetische Materialien »einzukapseln« oder aber diese Materialien selbst zu elektronisch reagierenden Kommunikationsträgern zu machen. Wir werden nochmals im letzten Kapitel darauf zurückkommen.

Auch gibt es Geräte für Kapazitätsupdates von iPods ähnlich dem Updaten bei Laptops und PCs. Der iWalk 88 z. B. ist ein externer Akku für die iPods classic, touch und nano sowie das iPhone. Er ermöglicht bis zu elf Stunden mehr Audio-, bis zu drei Stunden mehr Video- und bis zu zweieinhalb Stunden mehr Internetnutzung.

Im Oktober 2010 wurde iTunes in einer neuen Version vorgestellt. Zu ihr gehört nun ein musikzentriertes Social Network mit dem Namen Ping. Damit können die Nutzer gegenüber Bekannten und Freunden Musik präsentieren und bewerten. Diese neue Version von iTunes hat aber vor allem die Synchronisierung, die zur Verwaltung, zum Abspielen und zum Kauf der Musikstücke notwendig ist, nochmals vereinfacht. Zudem stellte Apple gleich drei neue iPod-Varianten vor. Vor allem der komplett neu entwickelte iPod nano ist nun vollkommen anders. Wie das iPhone und das iPad hat er nun auch einen berührungssensitiven Touch-Bildschirm. Das Gerät mißt nur 4 x 4 cm, ist 9 mm flach, wiegt lediglich 21 Gramm und wird in den Versionen 8 GB bzw. 16 GB angeboten. Dieser iPod nano kann an der Kleidung angeclipt werden und es gibt als Zubehör ein Armband, mit dem man das Gerät via Uhranzeige zur Armbanduhr machen kann. Eine Anwendung läßt sich beenden, indem man sie mit dem Finger nach rechts wegschiebt. Der ebenfalls überarbeitete iPod shuffle hat 2 GB Kapazität und wiegt 12,5 Gramm. Er hat nach wie vor das klassische Click-Wheel-Bedienrad, und es gibt ihn nun in fünf Farben. Am leistungsfähigsten allerdings ist der neue iPod touch. Sein neues sogenanntes Retina Display weist eine vierfach höhere Pixelauflösung wie sein Vorgänger auf. Zudem besitzt das Gerät zwei Kameras, so daß man wie beim iPhone 4 Videotelephonate führen kann, eine Funktion, die Apple Facetime nennt. Durch einen leistungsstärkeren Chip ist dieser neue iPod touch auch als mobile Spielkonsole geeignet. Es gibt Versionen mit 8, 32 und 64 GB. Bei angenommenen drei Minuten pro Song hat so das leistungsstärkste Gerät eine Kapazität von 800 Stunden, was 16 000 Songs entspricht oder, anders gesagt, einem Monat durchgehenden Dauerbetrieb. Konvergenzen also durchgehend: Der neue nano wirkt wie ein miniaturisiertes iPhone bzw. mit seinem schwarzen Rahmen wie eine quadratische Miniversion des iPad, der neue iPod touch nähert sich funktional den Nintendo-Spielkonsolen und das neue Tool Ping ist im Grunde eine Mischung aus Facebook und Twitter.

Ein weiteres Indiz für die gattungsbestimmende Attraktivität der iPods und der »Docking Stations« ist auch, daß sie – neben Espressomaschinen, Audio-Kompaktanlagen und Reisegepäck – zu den beliebtesten Prämien für die Neuwerbung eines Zeitungsabonnenten gehören. Selbst in anderen Produktgattungen entfaltet der iPod seine prägende Kraft. Das Steuergerät für Raumfunktionen Luxmate Ciria von Zumtobel, gestaltet 2010 von Matteo Thun, übernimmt die generelle Ästhetik des iPod classic, z. B. das Click Wheel, während das Gira Interface als Home Server App direkt für das iPhone entwickelt wurde. Sogar ein Akku-Handwärmer in der exakten Größe und Anmutung des iPod nano der 2. Generation belegt die Attraktivität und generelle Affinität von i-Produkten für an-

73. Ticker with iPod nano, 6.Generation, switcheasy, 2010.
74. AVA-N10LB with iPod nano, 6.Generation with wristwatch app, Elecom, 2010.
75. Rechargeable hand warmer HeizPod, Golfoholic, 2010.
76. U.S. Patent Click Wheel, Apple, 2006.

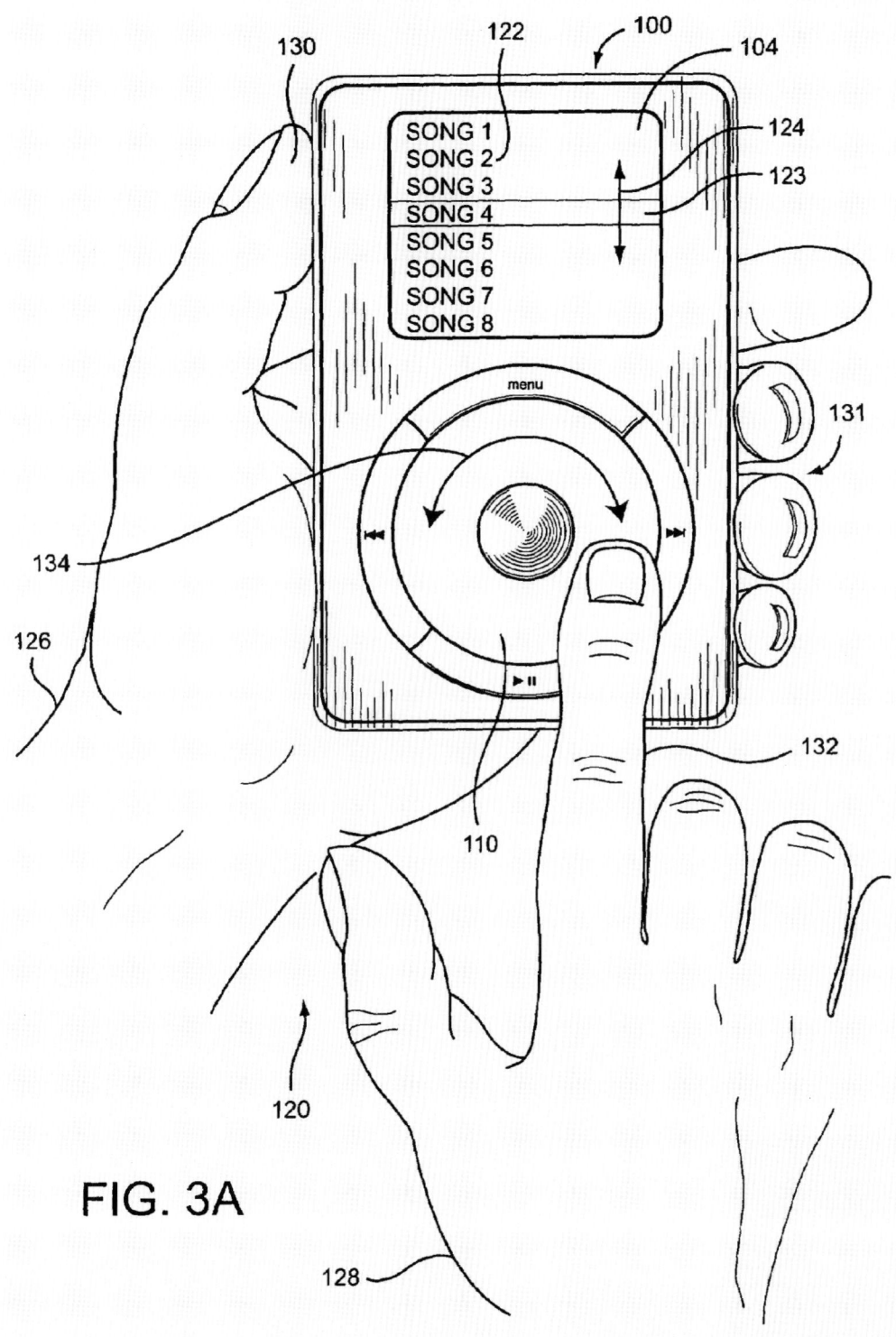

FIG. 3A

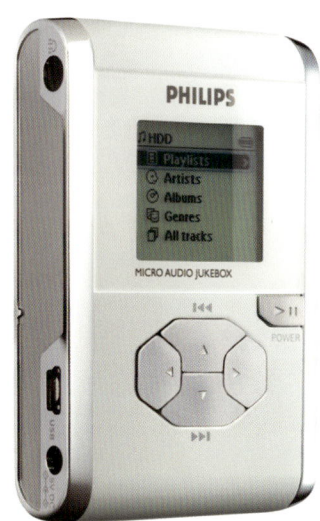
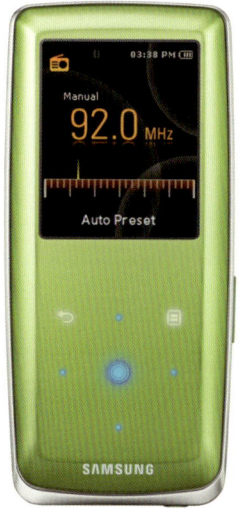
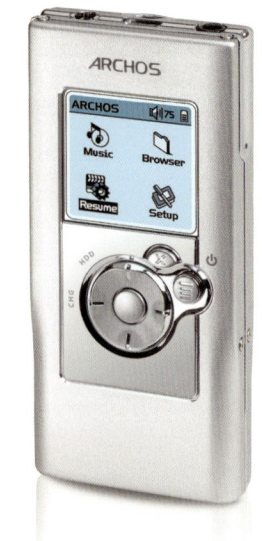
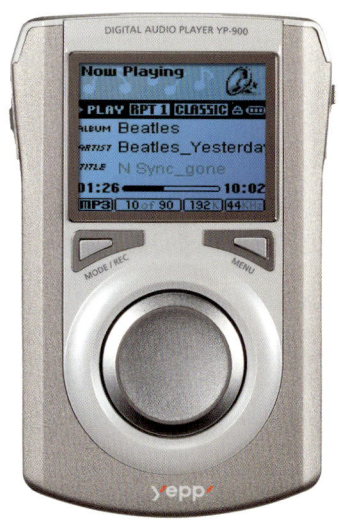

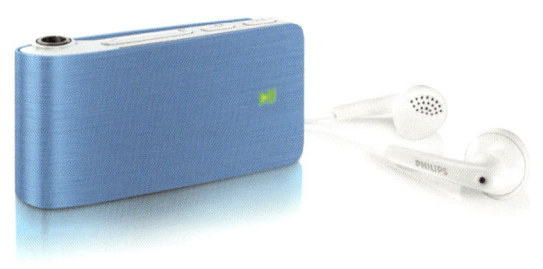
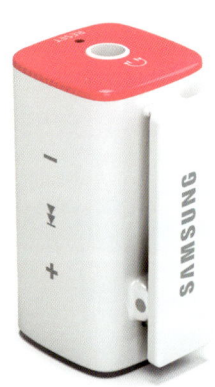
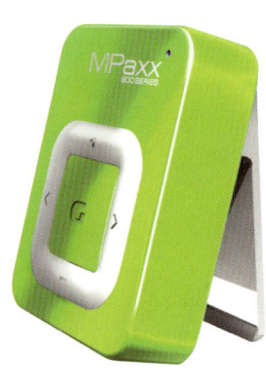

77. HDD075, Philips 2007.
78. YP-S3JA, Samsung, 2007.
79. gmini xs100, Archos, 2005.
80. yp 900, Samsung, 2006.
81. MPixx 1200, Grundig, 2010.
82. Fuze, SanDisk Sansa, 2007.
83. Zen V Plus, Creative, 2006.
84. zune HD, Microsoft, 2010.
85. zune 300, Microsoft, 2008.
86. N20, iriver, 2010.
87. YP S2 Q, Samsung, 2007.
88. Xemio-260, Lenco, 2010.
89. GoGear Clip, Philips, 2010.
90. TicToc, Samsung, 2010.
91. MPAXX 920, Grundig, 2010.
92. Sansa Clip+, SanDisk Sansa, 2009.
93. gmini xs202, Archos, 2005.

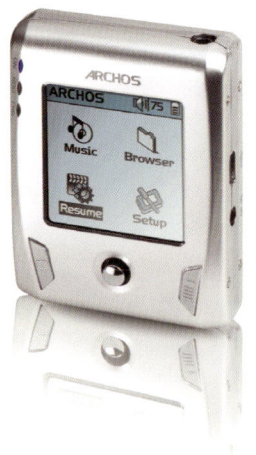

controller connected to the dock by a flat red conductive ribbon built into the watch pocket. The headphones retract into in a small pod attached to a belt loop.

While the textile shells in a sense protect the i-devices from damage internally, external textile »device extensions« for the i-family are available as well. The Austrian company Urban Tool, for example, produces very hip synthetic bags, holsters and T-shirts for i-products and other gadgets. The grooveRider is a functional shirt with a textile interface that allows one to control an iPod and its functions. The head phones are routed through a cable duct to the shoulders. The two collections by Urban Tool comprise around forty items and are named iWearables and iPortables. It is indeed a difference whether functional electronic elements are »encapsulated« in fabrics or synthetics, or whether these materials are turned into electronically responsive communication carriers. We will return to this in the final chapter.

There are also devices for upgrading the battery and storage capacity of iPods similar to the updates for laptops and PCs. The iWalk 88, for example, is an external battery for the iPods classic, touch and nano as well as the iPhone. It enables up to 11 hours more audio, up to 3 hours more video, and up to 2.5 hours more Internet use.

In October 2010 a new version of iTunes was released. It includes a social network for music called Ping through which users can follow their favorite artists and see what their friends and acquaintances are listening to, talking about and downloading. This new version of iTunes, however, mainly simplifies synchronization, which is necessary for the administration, playback, and purchase of music tracks. In addition, Apple presented three new iPod versions, especially the completely new iPod nano. Like the iPhone and iPad, it now also has a touchscreen. The device is a 4 x 4 cm square, has a thickness of 9 mm, only weighs 21 grams, and is offered in 8 GB and 16 GB versions. This iPod nano can be clipped onto any piece of clothing, and a bracelet is available as an accessory, turning the device into a wristwatch via its clock display. An application is closed by pushing it to the right with a finger. The also revised iPod shuffle has a 2 GB capacity and weighs in at 12.5 grams. It still has the classic click wheel and is now available in five colors. The new iPod touch, however, offers the highest level of performance. Its so-called retina display is a high-end screen with four times as many pixels as its predecessor. The device also has two cameras, enabling video telephony, as is the case with the iPhone 4, using Apple's Facetime app. Due to a better chip, this new iPod touch can also be used as a mobile games console. Versions with 8, 32, and 64 GB are available. Assuming 3 minutes per track, this high-performance device can store 800 hours of music, which equals 16 000 songs or, in other words, a full month of continuous operation. Convergence is hence continuous: the new nano looks like a miniaturized iPhone or with its black frame like a square mini version of the iPad; the new iPod touch functionally approaches the Nintendo games consoles, and Ping is basically a combination of Facebook and Twitter.

Another indication of the genre-defining attractiveness of the iPods and docking stations is the fact that, in addition to espresso machines, compact audio systems and travel luggage, they are one of the most popular bonuses given to new newspaper subscribers. Even in other product genres the iPod unfolds its formative power. The control device for room functions, Luxmate Ciria by Zumtobel designed 2010 by Matteo Thun, adopts the general aesthetics of the iPod classic, for example the click wheel, while the Gira Interface was directly developed for the iPhone as a home server app. Even a rechargeable hand warmer with the precise size and look of the second-generation iPod nano is proof for the attractiveness and general affinity of i-products for other genres. At the end of 2010, the company ADE Germany launched a digital kitchen scale with an iPod docking station. The products for building communications such as intercoms have meanwhile added docking stations to their product list as well. The company Jung, for example, offers such an intercom for wall mounting; it is as flat as an overdimensioned light switch

94. Jonathan Ive: iPod touch 4th generation, Apple, 2010.
95. iTunes prepaid card The Beatles, Apple, 2010.
96. iTunes prepaid cards, Apple, 2008.

97. iPod Tower, Lenco, 2009.
98. XW-NAC3, Pioneer, 2010.
99. XW-NAS5, Pioneer, 2010.
100. On Time, JBL, 2009.
101. Speakerball, Lenco, 2009.
102. Octiv 202, Altec Lansing, 2010.
103. Portable docking station Livespeaker, Digital Group Audio, 2010.
104. Go & Play, Kardon Harman, 2006.
105. Radiowecker ICF C05iP, Sony, 2010.
106. iYiYi, Tivoli, 2007.
107. icruiser, Avox, 2008.
108. Soulra SP400 with solar panel, Soulra, 2010.

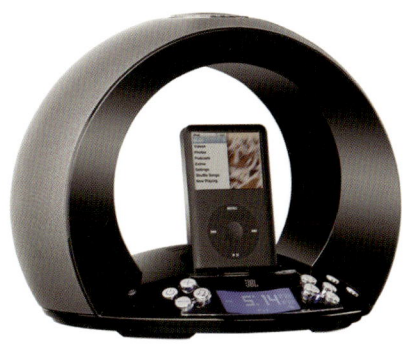

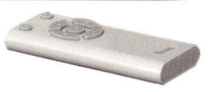

109. Philippe Starck: Zikmu, Parrot, 2009.
110. Jan Erik Lundberg: Genevalab XL, Geneva, 2010.
111. Docking station, Apple, 2006.
112. Charge N Fruits, Art in the city, 2010.
113. Breathe, Edifier, 2010.
114. Zeppelin, Bowers & Wilkins, 2010.
115. JBL On Stage III WM, JBL, 2010.
116. inMotion iM7, Altec Lansing, 2005.
117. ZonePlayer, Sonos, 2009.
118. TFT-2277, Lenco, 2009.
119. DK-AP8P, Sharp, 1984.
120. Popcube mini, Bergmann, 2010.

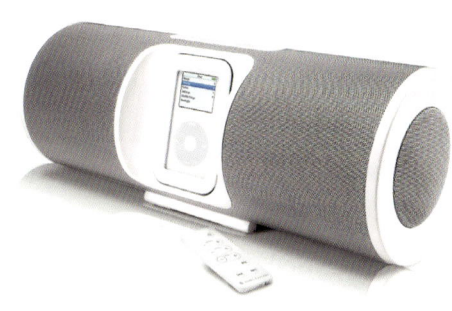

121. cuboDock, Sonoro, 2010.
122. Michael Young: MM04, Native Union, 2010.
123. E440, Sony, 2009.
124. Hohrizontal 51, Finite Elemente, 2010.
125. Soundstation, Tchibo, 2008.
126. BeoSound 8, Bang & Olufsen, 2011.
127. Creature III, JBL, 2007.
128. IPD, Lenco, 2009.
129. eklipse, Sonoro, 2009.
130. Cocoon, TrekStor, 2010.
131. iPal, Tivoli, 2004.
132. Sinus Sound Booster, Bergmann, 2010.

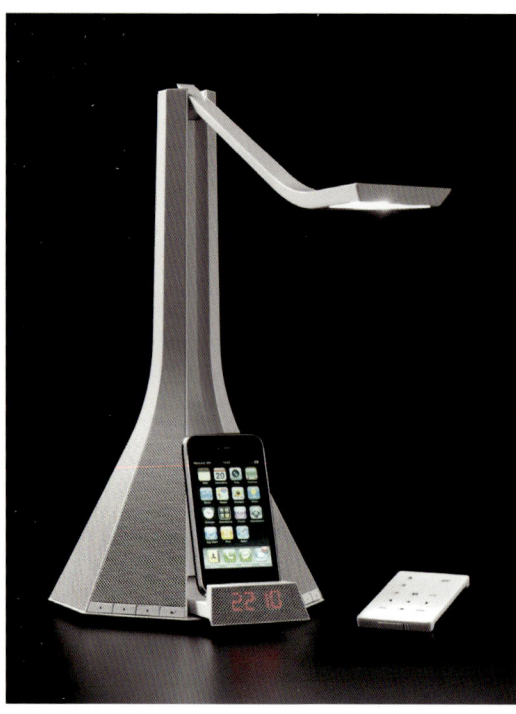 

dere Gattungen. Ende 2010 kam eine digitale Küchenwaage der Firma ADE Germany mit einer iPod »Docking Station« in den Handel. Auch Produkte für Gebäudekommunikation wie etwa Gegensprechanlagen haben inzwischen »Docking Stations« im Programm. So bietet das Unternehmen Jung eine solche zur Wandmontage an, die so flach ist wie ein überdimensionierter Lichtschalter und insofern an deren Produktcharakter anknüpft. Eine ironische Produktadaption des iPod classic bietet das Unternehmen Koziol an. Qed-Design aus Aschaffenburg hat für den Frühstückstisch den Eierbecher eiPott entworfen, eine Kunststoffschale in fünf Farben in der exakten Größe des MP3-Players. Sie hat zwei Vertiefungen in der Form des Displays und des Click Wheels. Die erste nimmt die Eierschalen, die zweite das gekochte Ei auf. Die Form dieses Eierbechers, seine Verpackung, vor allem aber sein Name, der onomatopoetisch, also lautmalerisch, den iPod herbeispielt, sind eine witzige Hommage an dieses Kultgerät für die Ohren. Apple selbst allerdings findet die klangliche Identität überhaupt nicht witzig und reagierte mit einer Abmahnung gegen Koziol. Das Landgericht ebenso wie das Oberlandesgericht Hamburg gaben Apple recht, weil sie eine markenrechtliche Verwechselungsgefahr erkannten. Diese humorresistente Reaktion von Apple hat Koziol süffisant gekontert. In einem neuen, nach dem Unterlassungsurteil verfaßten Beipackzettel heißt es in Anspielung auf das heroische Gallierdorf von Asterix und Obelix: »Wir befinden uns im Jahre 2010 n. Chr. Die ganze Welt huldigt einem angebissenen Apfel. Die ganze Welt? Nein, eine kleine Stadt im Odenwald hört nicht auf, dem Apfelkult Widerstand zu leisten. Und zwar mit einem unschuldigen Eierbecher. So klein, daß er fast übersehen worden wäre. Doch dann fühlte sich der Apfel belästigt von dem Winzling, der noch nicht mal Töne von sich geben kann. Ein hohes Gericht wurde angerufen. Und dann noch eines. Und noch eines. Bis endlich ein Erlaß erging, daß Eierbecher mit einem ei im Namen einfach verboten gehören. Woraus werden wir wohl morgen unser Ei löffeln? Und womit? Vielleicht nur noch mit einer kostenpflichtigen App?« Die Verpackung des eiPott zeigt im »Displayfeld« über dem exakt so geschriebenen Begriff die Silhouette eines »geköpften« Eis mit gezackter Schale. Auch dies ist neben der Schreibweise eine subtile Anspielung auf den angebissenen Apfel des Apple-Logos. Auf dem erwähnten Beipackzettel veränderte Koziol den Namen zu Pott und ergänzte die Ei-Silhouette mit einem liegenden Schalenstück. So ist es wohl unwahrscheinlicher geworden, daß irgendwann Hersteller von interaktivem Spielzeug, wie der »Furby«-Figur oder dem Computerdackel »Aibo«, ihr Erzeugnis »iPet« nennen werden, und ebenso wenig wird sich wohl die Bezeichnung »iFon« für die elektronisch geregelte Lautstärke von Konzertboxen durchsetzen. »Wer den Ärger hat, braucht für den sPott nicht zu sorgen.«

Im Oktober 2010 reagierte das deutsche Designhandelsunternehmen Ikarus auf diese, wie die Firma es nannte, »Ehe von Kunst, Kommerz und Komik«, indem sie jeder Warenbestellung ab 80 Euro zwei der nun nur noch Pott genannten Eierbecher gratis beipackte. In einem Rundbrief benannte das Unternehmen das Problem mit schöner Offenheit: »So kurios dieses Urteil ist, so gekonnt wurde hiermit die Freiheit des Designs, die ja auch eine künstlerische ist, ausgehebelt.«

Eine durchaus nicht ironisch gemeinte Adaption der weltweiten iPod-Euphorie ist ein 23stöckiges Wohn- und Bürohaus in schlanker Form und glatten, metallischen Oberflächen in der Form des iPods in Dubai, welches schon Anfang 2007 »iPad« genannt wurde. Entworfen hat dieses Hochhaus das ortsansässige Architekturbüro James Law Cybertecture International, die Baukosten sind auf 600 Millionen Euro geschätzt. Finanziert wird der Bau von der Firma Omniyat Properties. Durch eine leichte Schräglage und ein weit ausladendes Erdgeschoß wirkt das Gebäude wie ein iPod in seiner Ladestation.

133. Dante Donegani, Giovanni Lauda: lamp Diva, Rotaliana, 2009.
134. Lamp conof, Silver Seiko, 2009.
135. Cubes for iPod nano, switcheasy, 2010.
136. Colors for iPod touch, switcheasy, 2010.
137. ChocoShuffle for iPod shuffle, switcheasy, 2010.
138. Capsule thin for iPod nano, switcheasy, 2010.
139. iPod Socks, Apple, 2004.
140. Julia Oster: Freiwild Sleeve for iPhone and iPod touch, Freiwild, 2010.
141. Phone Garage, Frankie's Garage, 2010.
142. Marion Berger: Skin for iPod classic, elastica, 2005.

and in this regard relates to its product character. The company Koziol offers an ironic product adaptation of the iPod classic. Qed-Design from Aschaffenburg (Germany) has designed the egg cup eiPott for the breakfast table, a plastic shell with the exact dimensions of the MP3 player. It has two indentations in the shape of a display and the click wheel. The first takes the egg shells, the second the boiled egg. The shape of this egg cup, its packaging, but especially its name, which onomatopoetically conjures up the iPod, are a witty homage to this cult device for the ears. Apple, however, does not find this tonal identity funny at all and reacted by sending a written warning to Koziol. The district court and higher regional court Hamburg agreed with Apple because they recognized the danger of confusion of trademarks. This humorless reaction from Apple was complacently countered by Koziol. A new leaflet included in the package following the court injunction hints at the heroic village of the Gauls of Asterix and Obelix: »We are in the year 2010 AD. The whole world is paying homage to a bitten-into apple. The whole world? No, a small town in the Odenwald will not cease resisting the apple cult – with an innocent egg cup. So small that it could almost have been overlooked. But then the apple felt bothered by the dwarf that does not even emit sounds. It resorted to a higher court. And to another one. And another one. Until finally a decree was issued that egg cups with an ›ei‹ in their name simply have to be forbidden. What will we be eating our eggs out of tomorrow? And with what? Perhaps only with a chargeable app?« The packing of the eiPott presents the silhouette of a »decapitated« egg with a jagged shell in the »display field« above the term that is precisely spelled like this. This, too, alongside the spelling, is a subtle reference to the bitten-into apple of the Apple logo. On the aforementioned leaflet Koziol changed the name to sPott («Spott« in German means as much as »ridicule«) and added a piece of shell lying next to the egg silhouette. It has thus probably become more unlikely that at some point manufacturers of interactive toys such as the »Furby« character or the computer dachshund »Aibo« will name their products »iPet;« and neither will the term »iFon« for the electronically controlled volume of concert speakers prevail. »The laugh is always on the loser.«

In October 2010 the german design trading firm Ikarus reacted to this, as the company calls it, »marriage of art, commerce, and humor« by adding two of the now so-called Pott egg cups as a free gift for each order of over 80 euro. In a circular, the company commented on the problem with a fine openness: »As strange as this judgment may be, it skillfully leveraged the freedom of design, which is also an artistic freedom.«

In Dubai, a 23-floor apartment and office building with a lean design and smooth, metallic surfaces shaped like an iPod is absolutely not an ironically intended adaptation of the global iPod euphoria; it was named »iPad« in early 2007. The local architectural office James Law Cybertecture International designed this high-rise building; its construction cost is estimated to be 600 million euro. The company Omniyat Properties is financing the building. Due to a slight slant and strongly cantilevered ground floor, the building looks like an iPod in a charging station.

143. Qed Design: Pott, Koziol, 2009.
144. James Law Cybertecture: iPad skyscraper, Dubai, rendering, 2010.
145. James Law Cybertecture: iPad skyscraper, Dubai, model, 2011.
146. Matteo Thun: Luxmate Ciria, Zumtobel, 2010.
147. Music Center, Jung, 2010.

Der Kosmos der i-Geräte: Das iPhone und seine Parallel-, Peripherie- und Klonprodukte

Vielleicht mehr noch als die Benutzerführung der iPods durch Click Wheels hat das sensitive Touch-System der iPhone-Frontfläche die Produktlogik von digitalen mobilen Kommunikationsgeräten verändert. Das herausragende Merkmal des iPhone ist der berührungsempfindliche Bildschirm: Apple hat für das Gerät die Tastatur eliminiert. Um Telephonnummern einzugeben oder E-Mails zu schreiben, gibt es eine virtuelle Tastatur, die vom Nutzer mit dem Finger bedient wird. Es entstand damit ein komplett neues Verständnis von Gerätebedienung und Menüstruktur: keine Knöpfe, Tasten und Regler mehr, sondern eine Bildschirmfläche mit fest programmierten oder variabel programmierbaren »Fingerfeldern« für Organizer-Funktionen wie Kontakte, Aufgaben, Adressbuch, Kalender, Wetterberichte, Uhrzeit, Notizen, Taschenrechnerfunktion (das entsprechende Icon auf dem ersten iPhone war graphisch dem Braun-Taschenrechner von Dietrich Lubs nachempfunden), E-Mail, SMS, Freisprechfunktion, Telephon und Internetzugang mit kostenpflichtigen Online-Diensten. Ebenfalls verfügbare Funktionen sind ein UKW-Radio, ein MP3-Player sowie die Synchronisation von Outlook. Produktlogisch wie produktpsychologisch ist dies eine, wenn auch miniaturisierte, Adaption der Microsoft-»Windows«-Programme für PCs.

Das erste iPhone kam Anfang 2007 auf den Markt und wurde sofort euphorisch gefeiert. Es baute auf durchaus bereits vorher vorhandenen berührungssensitiven Touch-Oberflächen auf, nicht nur im Bereich der »consumer electronics«, sondern z.B. auch von Haushaltsgeräten wie Ceranfeldern oder Kühlschränken. Die Faszination des iPhone bestand von Anfang an darin, daß es im Grunde nur noch eine Fläche ist. Thomas Wagner schrieb zur Markteinführung des iPhone Anfang 2007 in der *Frankfurter Allgemeinen Zeitung*: »Ist das iPhone ausgeschaltet, so präsentiert sich der Touchscreen, der fast die gesamte Vorderseite einnimmt, als schwarz schimmerndes Nichts. Und genau so präsentiert sich auch das Design: als Projektionsfläche. Schon der Minimalismus des iPod zeichnete sich dadurch aus, daß sich auf ein milchig weißes Rechteck mit Bildschirm und kreisrundem Click Wheel problemlos ein neues Lebensgefühl projizieren ließ. Nicht das Objekt selbst ist wichtig, sondern die kulturelle Funktion, hinter der es verschwindet.«[20] Sowohl beim iPhone wie bei dem iPad sind es die berührungssensitiven Finger-Kommandos, die die Geräte dominieren, ja definieren. Der Apple-Chef Steve Jobs sagte bei der Erstpräsentation des iPhones im Januar 2007: »Wir sind alle mit dem ultimativen Zeigegerät geboren worden – unseren Fingern – und iPhone nutzt sie, um die revolutionärste Benutzeroberfläche seit der Maus zu schaffen.«[21]

Steve Jobs, der sich bei solchen öffentlichen Produktneuvorstellungen gerne in der Rolle eine Charismatikers gefällt, bezog sich zudem auf Michelangelos Deckenfresko »Die Erschaffung der Welt« (1536) in der Sixtinischen Kapelle in Rom, auf dem Gott mit ausgestrecktem Zeigefinger Adam berührt und ihm so Leben bzw. Bewußtsein implantiert: das iPhone also als »Bewußtseinstool«. Andererseits entbehrte dieser Ausflug in die Kunstgeschichte nicht einer gewissen Ironie, war doch dieses Fingermotiv von Michelangelo über Jahre hinweg der Display-Bildschoner aller Mobiltelephone von Nokia. Trotzdem ist Steve Jobs gegenwärtig wahrscheinlich der talentierteste Alchimist des Marketing überhaupt.

In der Tat wurde das iPhone von Anfang an als magisches Werkzeug verstanden, welches jedem, »der es kauft, (nicht nur) verspricht, ... unterwegs besser telephonieren, schneller Adressen finden oder leichter das Netz durchstöbern zu können.

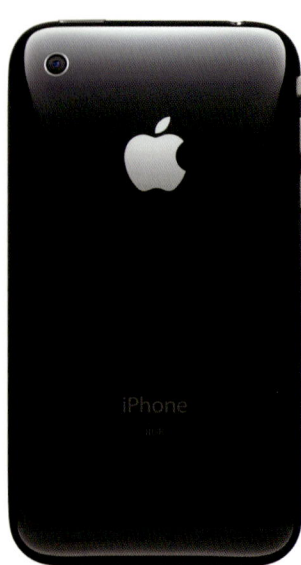

The cosmos of the i-devices: The iPhone and its peripheries, filiations, and adaptations

The touch-sensitive screen of the iPhone has perhaps changed the product logics of digital mobile communication devices even more than the user guidance of the iPod click wheel. The outstanding characteristic of the iPhone is the touch-sensitive display: Apple has eliminated the keyboard in this device. To enter phone numbers or write e-mails, a virtual keyboard is called up on the screen and is operated by a light touch of a finger. An entirely new understanding of device operation and menu structure was thus created: no buttons, keys or controls but a monitor surface with fixed programmed or variably programmable »finger fields« for organizing functions such as contacts, tasks, addresses, calendar, weather reports, time, notes, calculator functions (the appropriate icon on the first iPhone was graphically related to the Braun pocket calculator by Dietrich Lubs), e-mail, SMS, hands-free function, phone, and Internet access with fee-based online services. Other available functions are an FM radio, MP3 player, and synchronization with Outlook. In terms of product logics as well as product psychology, this is absolutely an adaptation – although miniaturized – of Microsoft's »Windows« programs for PCs.

The first iPhone was launched in early 2007 and was immediately celebrated with euphoria. It built on previously existing touch-sensitive screens, not only in the consumer electronics segment but also of household appliances such as Ceran stovetops or refrigerators. The fascination of the iPhone from the beginning was that it basically is just a surface. Thomas Wagner wrote in *Frankfurter Allgemeine Zeitung* in the beginning of 2007 on the market introduction of the iPhone: »When the iPhone is switched off the touchscreen, which occupies almost the entire front of the device, presents itself as a shimmering black nothing. And the design presents itself just like that: as a projection surface.

The minimalism of the iPod already stood out due to the fact that a new lifestyle could be projected onto a milky-white rectangle with a screen and circular click wheel. It is not the object itself but the cultural function behind which it disappears that is important.«[20] For both the iPhone and the iPad, the touch-sensitive finger commands dominate, even define, the devices. Apple CEO Steve Jobs said during the first presentation of the iPhone in January 2007: »We were all born with the ultimate pointing device – our fingers – and iPhone uses them to create the most revolutionary user interface since the invention of the mouse.«[21]

Steve Jobs, who likes to play the role of a charismatic during such public product introductions, also referred to Michelangelo's ceiling fresco in the Sistine Chapel entitled »Creation of Adam« (1536) in which God is reaching out to touch Adam with a pointed index finger to implant life or consciousness in him: the iPhone as a »consciousness tool.« On the other hand, this excursion into art history did not lack a certain irony since this finger motif of Michelangelo had been the display screensaver of all of Nokia's mobile phones for years. Still, Steve Jobs is probably the most talented alchemist of marketing in the world today.

In fact, from the beginning the iPhone was understood as a magical tool that promises to everybody »... who buys it (not only) to be able to have better telephony, find addresses more quickly and browse through the web more easily.« Its message is: »I am a magic mirror, and if you touch me then you can touch the whole world. You just have to put your fingers on something and it is at your service.«[22] The fascination of use is not only in the touching of individual icons; items on the screen can also be »stretched«, meaning the image can be increased in size by spreading the index finger and thumb on the

148. Jonathan Ive: iPhone 1st generation, Apple, 2007.
149. Jonathan Ive: iPhone 2nd generation, Apple, 2009.
150. Steve Jobs demonstrates the iPhone 4 in San Francisco, 2010.

Seine Botschaft lautet: Ich bin ein Zauberspiegel, und wenn du mich berührst, dann berührst du die ganze Welt. Du brauchst nur die Finger auf etwas zu legen, und schon ist es dir zu Diensten.«[22] Dabei besteht die Faszination der Benutzung nicht nur im Antippen eines einzelnen Feldes, sondern man kann die Bildfläche »aufziehen«, d. h. den Bildausschnitt vergrößern, indem man Zeigefinger und Daumen auf der sensitiven Glasoberfläche auseinanderspreizt. Damit wird der phototechnische Vorgang des Zoomens durch ein Teleobjektiv überführt in eine anthropogene Handlung: eine durchaus magische Verwandlung. Und natürlich kann man das Bild aufgrund des Rotationsdisplays von der Vertikalen in die Horizontale drehen.

Auf dem gläsernen Display des iPhone sind bei aller Aktualität der Geräte und seiner avantgardistischen Benutzerführung die centgroßen Icons erstaunlich konventionell, manchmal nostalgisch. So zeigt das Icon für die Funktion YouTube einen 1950er Jahre-Fernseher, das Piktogramm für Zeitmessung eine kreisrunde Wanduhr der gleichen Zeit, jenes für »Sprachmemos« ein maiskolbenförmiges Standmikrophon der 1940er Jahre und die Rechnerfunktion den Ausschnitt eines analogen Taschenrechners. Die Telephonfunktion wird mit dem »Knochenhörer« eines Schnurtelephons symbolisiert, die Mail-Funktion mit einem Briefumschlag und die iPod-Funktion mit einem abstrahierten Logo des iPod-Gerätes selbst. Geradezu nostalgisch wird die Funktion »Einstellungen« mit drei ineinandergreifenden Zahnrädern visualisiert, die nicht nur das Zeitalter der Mechanik im frühen 19. Jahrhundert herbeispielen, sondern auch die Bildsprache des Films *Modern Times* (1936) von Charles Spencer Chaplin. Samsung symbolisiert die Funktion »Video-Player« mit der Graphik einer 35-mm-Filmkamera mit aufgesetzten Filmrollen, während das Icon für die »Alarm«-Funktion ein runder nostalgischer Küchenwecker mit zwei Alarmschellen ist. Dieses Verhaftetsein mit vordigitalen Symbole ist ein deutlicher Hinweis auf die generelle Unanschaulichkeit digitaler Prozesse: Offenbar müssen sie immer noch oder können nur analog repräsentiert werden.[23]

Ende Juni 2010 kam die vierte iPhone-Generation auf den Markt. Neben dem Display ist nun auch die Rückseite aus Glas, und statt der abgerundeten Kanten verläuft ein Stahlband um das Gerät, welches nicht nur als Aufprallschutz, sondern gleichzeitig auch als Antenne dient. Beim Display ist die Auflösung der Pixel so groß, daß sie mit bloßem Auge nicht mehr zu unterscheiden sind. Auch die Kamera auf der iPhone-Rückseite ist erneut leistungsstärker. Erstmals ist Videotelephonie integriert: Man kann seinen Gesprächspartner, sofern er ein gleiches Gerät benutzt, sehen. Diese neue Funktion nennt Apple facetime. Zudem können erstmals mehrere Apps gleichzeitig und nicht nur nacheinander aufgerufen werden. Apple bezeichnet dies als multitasking.

Aber selbst ein so renommiertes Unternehmen wie Apple kann auch einmal Qualitätsprobleme bekommen. So verursacht der Antennen-Rahmen des iPhone 4 ab und zu Empfangsprobleme. Apple hat deshalb allen bisherigen und künftigen Käufern gratis eine Kunststoffhülle angeboten, die dieses auch Antennagate genannte Problem verhindert. Durchaus von Sympathie getragene, gleichwohl ironische Kommentare und Bildmontagen im Internet blieben nicht aus. Manche der Photomontagen haben eine geradezu frappierende Chuzpe. So zeigt eine Bildmontage das iPhone 4 mit fingerlanger Stabantenne und eine weitere das Gerät im Profil, an dessen Rückseite als Griff ein »Knochenhörer« aus den fernen Zeiten der Schnurtelephone montiert ist, eine dritte schließlich zwei Stabantennen, die an den »Weltempfänger T 1000« (1963) von Braun erinnern.

Der neueste Schachzug des Unternehmens besteht in dem Plan, in dem für das Frühjahr 2011 angekündigten iPhone 5 die SIM-Karte direkt einzubauen. Die Kunden würden dann ihr iPhone nur noch direkt bei Apple kaufen und ihren Netzbetreiber und Tarif im Apple Store auswählen. Dies hätte für die Netzbetreiber weitreichende Konsequenzen: Sie würden ihre Mobilfunkkapazitäten für das iPhone nur noch an Apple verkaufen. Bisher zwingen die Anbieter die Kunden dazu, teurere 24-Monatstarife abzuschließen. Damit werden die Geräte gewissermaßen subventioniert und für einen Minipreis angeboten. Der Verlust der Kundenbindung wäre für die Netzbetreiber schmerzlich. Apple würde einen weiteren Teil der Wertschöpfungskette im Mobilfunk an sich ziehen. Dieser Affront gegen die Netzbetreiber wäre bei einem gegenwärtigen Apple-Marktanteil von 17 % weltweit bei allen verkauften Smartphones auch durchsetzbar. Die Umgehung bzw. Aushebelung der bisherigen Wertschöpfungsaufteilung ist erneut eine disruptive Technologie, die die Kette von Gerätekosten, Netzkosten, Nutzungsgebühren und Sonderrabatten neu konfiguriert: eine ebenso frappierende wie marktsubversive Idee. Ich habe eine solche Marktstrategie bereits 1989 als »Produktfeldpiraterie« beschrieben. Auch diese jüngste Unternehmensstrategie ist von jener Chuzpe gekennzeichnet, für die Steve Jobs berühmt und berüchtigt ist.

Die Aktualität des iPhones ist bewußt zeitlich limitiert. Ähnlich wie bei Computern und deren Programmen gibt es bei allen i-Geräten »Generationen«: Sie werden »upgedatet«. Dies entspricht den saisonalen Strategien der Mode, was nachdrücklich durch individualisierte Geräteschalen unterstrichen wird. Es gibt Dutzende von iPhone-Outfits, von konservativ bis stylish, von hip bis zum Gothic style. Angeboten werden Gehäuse in grellen Popfarben, mit Reptilleder- oder Tigerfellanmutung, zweifarbig abgesetzte Geräte, Gehäuse im Holzlook, voll verchromte Oberflächen oder romantisch-plüschige Dekorapplikationen. Diese umfangreiche Diversifizierung entspricht den bereits beim iPod beschriebenen Strategien der »customization« und der »Swatchisierung«. Bereits vor zehn, fünfzehn Jahren gab es individualisierte Gehäuse für Mobiltelephone, z. B. bei Nokia, die selbst noch »Ereignisornamentik« anbot: etwa Sondereditionen für Weihnachten oder Ostern.[24] Solche Gehäusestrategien nobilitieren die Mobiltelephone und verleihen ihnen eine Aura, die die normalen Großseriengeräte nicht haben (können). Das Unternehmen GroveMade bietet eine iPhone-Schale aus Bambus an und die Firma HTC bei seinem Modell Tatoo sogar die Möglichkeit, wechselnde Coverschalen selbst zu gestal-

151. Anon.: Antenna fix, xayni, 2010.
152. Paul Strauss: Antenna fix, 2010.
153. Peter J. Morgan: Antenna fix, 2010.
154. Jonathan Ive: iPhone 4, Apple, 2010.

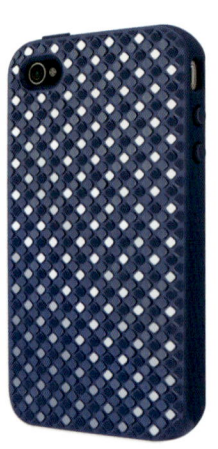

155. Clemens Burkert: Soay, Pack & Smooch, 2010.
156. Ilka Brand: Braid black coffee, Lapàporter, 2010.
157. Bamboo, Grove Made, 2010.
158. DesignSkin, DeinDesign, 2010.
159. Odyssey, switcheasy, 2010.
160. Body Hue, Belkin, 2007.
161. Reptile for iPhone 4, switcheasy, 2010.
162. Glitzblue, switcheasy, 2010.
163. Apps social Media, 2010.

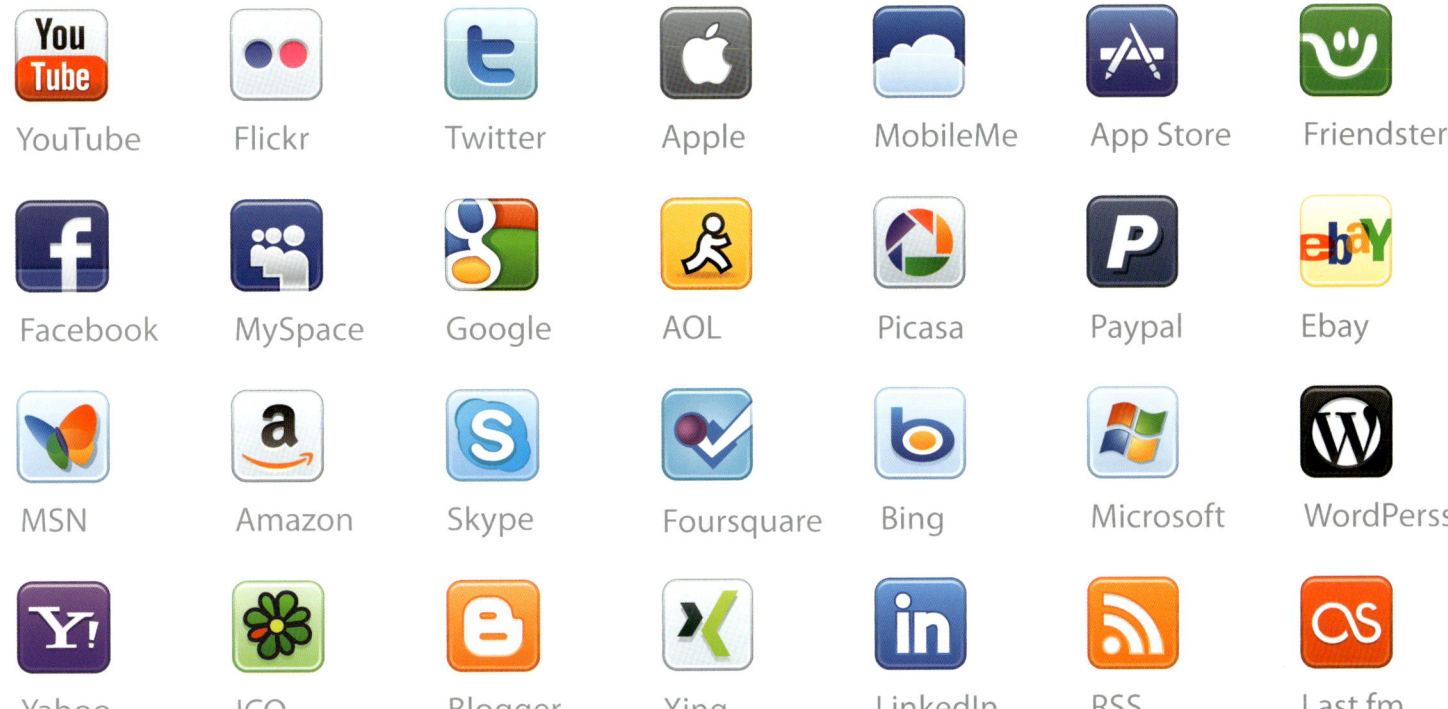

touch-sensitive glass surface. The photographic process of zooming in with a telephoto lens is transported into an anthropogenic act: an absolutely magical transformation. And, of course, due to the built in accelerometer, the image can be rotated from vertical to horizontal.

On the glassy display of the iPhone – despite all topicality of the devices and the avant-garde user guidance – the dime-size icons are surprisingly conventional, at times even nostalgic. The icon for YouTube is a 1950s television set, the pictogram for time is a circular wall clock from the same era, the one for voice memos is a corn-cob shaped microphone from the 1940s, and the calculator is represented by an icon of an analog pocket calculator. The phone function is symbolized by the »bone receiver« of a hardwired landline phone, the mail function by an envelope, and the iPod function by an abstract logo of the iPod itself. The »settings« function is visualized in an almost nostalgic way with three interconnected toothed wheels that not only allude to the mechanical era of the early 19th century but also the pictorial language of Charles Spencer Chaplin's movie *Modern Times* (1936). Samsung symbolizes the »video player« function with the icon of a 35mm film camera with attached film reels, while the icon for the »alarm« function is a nostalgic, round clock with two alarm bells. This bondage to pre-digital symbols is a clear hint at the general inability to illustrate digital processes: obviously, they still have to or can only be represented through analog means and methods.[23]

The fourth generation of the iPhone was launched at the end of June 2010. In addition to the display, the back is now made of glass, and instead of the rounded edges, a steel band embraces the device, serving not only as impact protection but also as an antenna. The pixel resolution of the display is so great that the individual pixels can no longer be differentiated with the naked eye. The resolution of the camera on the back of the iPhone has again been increased as well. For the first time, video telephony has been integrated into the device: users now have the ability to see the person they are calling, provided he or she is also using an iPhone. Apple calls this new function facetime. In addition, several apps can now be accessed simultaneously instead of one at a time. Apple calls this multitasking.

But even companies as renowned as Apple can run into quality problems. The antenna is built into the frame of the iPhone 4, and this sometimes causes reception problems. Apple has therefore offered a free plastic shell to all past and future buyers that eliminates this problem, which has been referred to as Antennagate. Ironic comments and montages on the Internet, which are at the same time absolutely driven by sympathy, naturally accompany the problem – some of the photomontages have an almost striking chutzpah. One photo montage shows the iPhone4 with a finger-length antenna, and another presents a profile of the device with a »bone-shaped receiver« from the long past era of hardwired telephones mounted to its back as a handle, a third one combines the iPhone with a double antenna in the manner of the »worldreceiver T 1000« (1963) by Braun.

The company's most recent move is the plan to directly install SIM cards in the iPhone 5, which was announced for spring 2011. Customers would then only be able to purchase iPhones directly from Apple and choose a network operator and plan in the Apple Store. For the network operators, this would have far-reaching consequences: they would sell their mobile services for the iPhone exclusively to Apple. So far, providers have been forcing customers to sign up for expensive 24-month plans. In a way, the devices are subsidized and offered at a minimum price. The loss of customer retention would be painful for the network operators. Apple would attract a large portion of

ten. Und schon in den 1980er Jahren gab es diverse Kleinunternehmen, die per Handzettel dafür warben, ein Mobiltelephon durch Airbrush-Technik zu individualisieren, so wie dies ein Jahrzehnt vorher bei manchen Automobilen en vogue war. Zur »customization« gehören auch jene kleinen Figuren, die als baumelnde Schmuckapplikationen, die an den meisten Mobiltelephonen angebracht werden können. In den 1990er Jahren waren vor allem in Japan und Korea für fast alle jugendlichen Nutzer diese Mini-»Love dolls« ein »must have«-Produkt.

Auch das iPhone hat wie der iPod seine Gattung verändert. Mobiltelephone mit berührungssensitiven Multitouch-Displays haben inzwischen so gut wie alle Mitbewerber im Programm: Acer, BlackBerry®, Google, HTC, LG, Motorola, Nokia, Palm, Pearl, Philips, Samsung und Sony Ericsson. Von diesen Firmen waren Mitte 2010 zusammen ca. 20 Multitouch-Geräte im Handel. Und so hat sich neben der Bezeichnung »Kamera-Handy« der Begriff »Touch-Handy« etabliert. Allerdings laufen mindestens die Hälfte davon mit dem Google-Betriebssystem Android, das dem Betriebssystem iPhone OS ernsthaft Konkurrenz macht. Derzeit aktiviert Google täglich 100 000 Android-Smartphones. Im ersten Halbjahr 2010 wurden mehr Android-Geräte als iPhones verkauft. Um den Erfolg seines Betriebssystems zu sichern, hat Google sein Smartphone Nexus One vom Markt genommen, weil es in allen Netzen funktionierte und z.B. durch Internet-Telephonie das Geschäft der Netzbetreiber hätte schädigen können. So verkauft Google Android hauptsächlich an andere Hersteller. Ein Nachfolgemodell zum Nexus One, das Nexus S, kam Anfang 2011 auf den Markt.

Aber es gibt auch Hybrid-Handys, die zwar ein Touchscreen-Display haben, aber auf eine umfangreiche Tastatur nicht verzichten wollen. So wird das Mobiltelephon GT 350 von LG auch als Slider bezeichnet, weil es unter der Touch-Oberfläche eine an der Längsseite herausschiebbare Tastatur hat. Unter Konvergenzgesichtspunkten ist dieses Gerät ein Hybrid zwischen BlackBerry® und iPhone. Auch der BlackBerry®-Hersteller RIM® schickt ein Smartphone mit Touchscreen und Tastatur ins Rennen gegen das iPhone und die Android-Geräte. Zwar ist der Konzern für die Manager-Kundschaft immer noch die Nummer zwei bei den Smartphones nach Nokia. Doch das iPhone holt atemberaubend schnell auf. Das Torch 9800 von RIM® hat hinter dem Touchscreen eine Tastatur, die nach unten ausgefahren werden kann. Das Unternehmen verspricht zudem eine verbesserte Benutzeroberfläche durch sein neues Betriebssystem 6. Im Grunde genommen aber sind solche Hybrid-Geräte Rückzugsgefechte, wenn auch auf hohem Niveau. Die zunehmende Dominanz der reinen Smartphones wird dadurch weder verhindert noch auch nur verlangsamt.

Aber jenseits aller betriebswirtschaftlichen Zahlen markieren die Smartphones mit dem iPhone an der Spitze einen nicht mehr kündbaren demokratischen Zugriff auf im Grunde alle Informationen der Welt: Und diese Informationen sind internetbedingt prinzipiell nicht kontrollierbar. Neben aller unterhaltungselektronischen Prominenz macht dies die Geräte in mancher Hinsicht durchaus subversiv, wie ja überhaupt die Wirkungsgeschichte der Massenkommunikation von Anfang an ambivalent gewesen ist. Vom Volksempfänger bis zum Kinosaal, vom privaten TV bis zum Internet halten sich Aufklärung und Verdummung die Waage. Im übrigen begünstigt das iPhone in vielen Schulklassen auch Betrügereien, denn nicht wenige Schüler »spicken« bei Klassenarbeiten im mobilen Internet. Die meisten Lehrer nehmen dies offenbar achselzuckend bis billigend zur Kenntnis, da so das Notenniveau der Klassenarbeiten steigt und Wiederholungen dieser Tests vermieden werden. So hat das iPhone auch einen neuen Techno-Zynismus in die Pädagogik gebracht und schulische Sozialisationsregeln erodieren lassen

Aber es bleibt die Frage: Hat bei den Smartphones das Telephonieren nur noch eine Alibifunktion bei Hunderten von weiteren Funktionen? Allerdings erwerben durchaus manche Käufer ein Smartphone schon mit der Absicht, mit ihm wirklich nur zu telephonieren. Sie bezahlen also für etwas, das sie nicht nutzen. In den 1980er Jahren nannte dies die Marketingtheorie »demonstrativen Verbrauch«. Hier mag man eher von »elektronischem Balzverhalten« reden.

164. Palm Pilot, Palm, 1996.
165. Palm Pre, Palm, 2009.
166. Legend, HTC, 2010.
167. GD510, LG, 2007.
168. Nexus One, Google, 2010.
169. Trophy, HTC, 2010.
170. Bravo, Motorola.
171. X3, Nokia, 2010.
172. I5700 Galaxy Spica, Samsung, 2008.
173. Xperia X8, Sony Ericsson, 2010.
174. i900, Samsung, 2007.

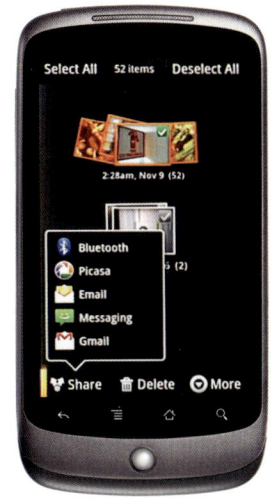

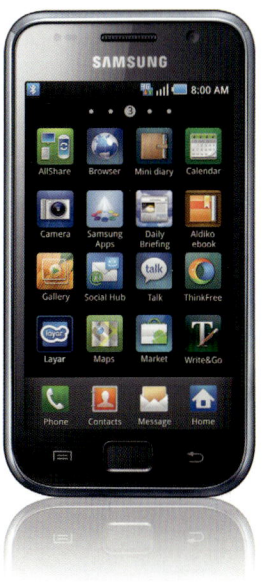

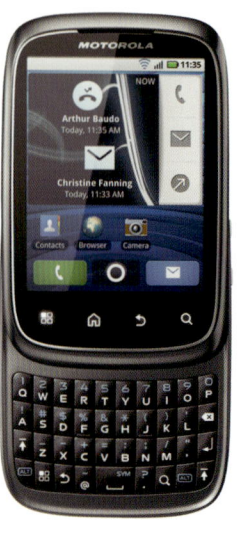

the value chain in the mobile segment. This affront to the network operators would also be feasible considering Apple's current market share of 17 % worldwide with regard to all smartphones sold. The bypassing and undermining of value chain division is again a disruptive technology, which reconfigures the chain of device costs, network costs, usage fees and special discounts; it is an idea that is as striking as it its market subversive. I already described such a market strategy in 1989 as »product field piracy.« This most recent corporate strategy is also characterized by the chutzpah for which Steve Jobs is both famous and infamous.

The topicality of the iPhone was consciously limited with respect to time. Similar to computers and their programs, there are »generations« for all i-devices: they are updated. This complies with the seasonal strategies of fashion, which is expressly underscored by individualized device shells. There are dozens of iPhone outfits, from conservative to stylish, from hip to Gothic style. Casings are offered in bright pop colors, with reptile leather or tiger fur design, bicolor casings, and casings in wood design, fully chrome-plated surfaces or romantically plush décor applications. This comprehensive diversification equals the strategies of customization and »Swatchization« that was already described for the iPod. Ten, fifteen years ago there already were individualized casings for mobile phones, for example from Nokia, that offered »event decorating« such as special editions for Christmas or Easter.[24] Such casing strategies ennobled the mobile phones and provided them with an aura that the regular large series devices do not (and cannot) have. The company Grove-Made offers an iPhone shell with an individual design made of bamboo, and HTC even offers users the possibility to design changing cover shells for its Tatoo model. Even during the 1980s there were many small companies that used flyers to advertise the individualization of mobile phones through airbrushing, as was *en vogue* one decade earlier for some automobiles. The small figures that can be attached to most mobile phones as dangling decorative applications are also part of the customization. In the 1990s these mini »love dolls« were a must-have product for almost all young users, especially in Japan and Korea.

The iPhone, like the iPod, has radically changed its product genre. Mobile phones with touch-sensitive, multi-touch displays have meanwhile become part of the product lineup of all competitors: Acer, BlackBerry®, Google, HTC, LG, Motorola, Nokia, Palm, Pearl, Samsung, and Sony Ericsson. In mid-2010 a total of approximately 20 multi-touch devices were available from these companies combined. Alongside the term »camera phone« the term »touch phone« has become an established term. However, at least half of them run the Google operating system Android, which offers serious competition to the iPhone operating system. Currently, Google is activating 100 000 Android smartphones per day. In the first quarter of 2010 more Android phones were sold than iPhones. In order to secure the success of its operating system, Google has discontinued its smartphone Nexus One because it functioned in all networks and, for example, could have damaged the business of the network operators through Internet telephony.

Hence, Google is selling Android mainly to other companies. A successor model to the Nexus One, the Nexus S, was launched in the beginning of 2011.

However, there are also hybrids that have touch-screen displays but also include a separate keyboard. The mobile phone GT 350 of LG is also called slider because it has a slide out keyboard under the touch display. Under aspects of convergence, this device is a hybrid between a BlackBerry® and an iPhone. Research In Motion® (RIM®), the creator of the BlackBerry®, is also sending a smartphone with touchscreen and keyboard into the competition against the iPhone and the Android devices. RIM® is still No. 2 among the management clients with regard to smartphones after Nokia, but the iPhone is catching up with breathtaking speed. The new Torch 9800 by RIM® has a keyboard behind the touchscreen that can be slid out towards the bottom. The company also promises an improved user interface due to its new operating system BlackBerry® 6. Basically, however, such hybrid devices are rearguard actions, although at a high level. The increasing dominance of the pure smartphones is neither prevented nor even slowed down through them.

But beyond all economic numbers, the smartphones headed by the iPhone mark the interminable democratic access to basically all information worldwide, and this information can in principle no longer be controlled due to the Internet. In addition to all prominence in entertainment electronics, this makes the devices absolutely subversive in some respects, just as the history of the influence of mass communications has been ambivalent from the beginning. From the public radio to movie theaters, from private TV to the Internet – education and stultification have been kept in balance. Apart from that, the iPhone also enables cheating in many school classrooms because quite a few students »peak« into the mobile Internet during tests. Most teachers obviously accept it with a shrug of their shoulders or willingly tolerate this cheating since it raises the grade level of the tests and avoids having to repeat them. The iPhone thus has also brought a new techno-cynicism to education and led to the erosion of socialization rules in schools.

However, the question remains: is making phone calls only taking on the function of an alibi among hundreds of other functions in the case of smartphones? Some buyers may actually purchase a smartphone with the intention of actually just making phone calls with it. Hence, they end up paying for capabilities they do not use. In the 1980s, marketing theory called this phenomenon »demonstrative consumption.« Here, however, one may rather speak of »electronic courtship behavior.«

175. E7, Nokia, 2010.
176. Flipout, Motorola.
177. GT350, LG, 2010.
178. X5, Nokia, 2010.
179. BlackBerry® Torch™ 9800 smartphone, RIM®, 2010.
180. BlackBerry® Torch™ 8800 smartphone, RIM®, 2007.
181. Spice, Motorola, 2010.
182. AS-iP70, Kenwood, 2010.

Der Kosmos der i-Geräte: Das iPad und seine Parallel-, Peripherie- und Klonprodukte

Das iPad ist ein flacher Tablet-Computer, auch »Tablet« genannt, mit einem in der Diagonale knapp 30 cm messenden Multitouch-Display, welches mit LEDs hinterleuchtet ist und im April 2010 in den USA auf den Markt kam. Dieses extrem schlanke Lifestyle-Gerät, graziler als jeder Laptop oder jedes Netbook, erlaubt Surfen im Internet, E-Mails zu empfangen und zu senden, Videos, Bilder und TV-Programme zu betrachten, Videospiele zu spielen und Musik zu hören. Fast alle der ca. 200 000 Programme aus den App Stores können darauf laufen, auch jene, die bereits für den iPod touch oder das iPhone gekauft wurden. Photographien von PCs und Digitalkameras sind übertragbar, zu archivieren und zu arrangieren.

Vor allem aber soll dieses Graphiktablet das E-Reading, das Geschäft mit den elektronischen Ausgaben von Tageszeitungen und Magazinen ankurbeln. Denn nun sind nicht nur einfach übertragene Printseiten mit statischen Bildern möglich, sondern auch in die Texte eingeklinkte Videos, Verlinkungen mit Zusatzinformationen wie gefilmten Hintergrundinterviews, natürlich aber auch bewegte Commercials. Außerdem lassen sich per Fingerbewegung auf der Glasfläche die Buch- bzw. Zeitungsseiten »umblättern«: So simuliert die digitale Technik Gewohnheiten der analogen Welt.

Da hat Apple auch aus eigenen und Fehlern anderer Anbieter gelernt. Schon 1991 brachte Sony das Gerät Data Discman auf den Markt, das die CD-Technik für E-Reading in sechs, sieben Sprachen nutzte. Apples Message Pad Newton (ab 1997) war ein frühes elektronisches Buch-Lesegerät (ab Mitte 1994), dessen Display auch per Hand beschrieben werden konnte. Es kam im Grunde zehn Jahre zu früh auf den Markt und hatte so niedrige Absatzzahlen, daß Steve Jobs nach seiner Rückkehr 1997 zu Apple die Produktion einstellen ließ. Ein Grund dafür war auch, daß nach seiner Markteinführung 1996 der vergleichbare Palm Pilot sich innerhalb von 18 Monaten über eine Million Mal verkaufte. 2007 stellte der Online-Händler Amazon in den USA das elektronische Lesegerät Kindle vor. Auch dies brachte wirtschaftlich zunächst bei weitem nicht den Erfolg, den die Verlagsbranche sich erhofft hatte. Das lag erstens an der relativ bescheidenen Taschenbuchgröße dieses E-Books, zweitens an dem Zwang, Buchseiten nur senkrecht scrollen zu können und drittens daran, daß die Bücher schlicht nur eingescannt waren. Ein wirklicher Medienvorteil gegenüber gedruckten Büchern war kaum vorhanden. (Seit 2009 gibt es im App Store die Applikation Kindle for iPhon – ein Vorgang nicht ohne Ironie). Das iPad hat alle diese Einschränkungen nicht: es ist zwar nur wenig größer, aber eben multimediafähig, und die eingescannten Bücher sind, wie gesagt, waagrecht umzublättern. Immerhin setzt Amazon weiterhin unverdrossen auf die Vermarktung seiner elektronischen Bücher: Ende August 2010 brachte das Unternehmen eine neue Version des Kindle in den Handel. Auch andere Hersteller geben sich neuerdings wieder optimistisch. So bietet Odys sein Media-Book Scout mit einem Minidisplay von gerade einmal 13 cm an, das nicht nur ein Lesegerät ist, sondern auch Musik-, Video- und Photodateien speichern kann. Die größte Buchhandelskette Amerikas, Barnes & Noble, bietet als elektronisches Lesegerät ihren Nook an, der sich allerdings bisher gegen Kindle und iPad nicht wirksam positionieren konnte. Insgesamt jedoch steigen die Verkaufszahlen der E-Books, wohl kaum zufällig nach der Markteinführung des iPads, gegenwärtig rasant. Je nach Speichergröße passen mehrere tausend Bücher auf ein E-Book. Zur Zeit sind von ca. 1,2 Millionen deutschsprachiger Titel rund 70 000 als E-Book verfügbar. Im zweiten Quartal 2010 hat Amazon erstmals mehr digitale als gedruckte Bücher verkauft. Schon wird prognostiziert, daß spätestens 2012 ein Viertel aller amerikanischen Bücher als E-Bücher verkauft werden. Auch Sony hat E-Books

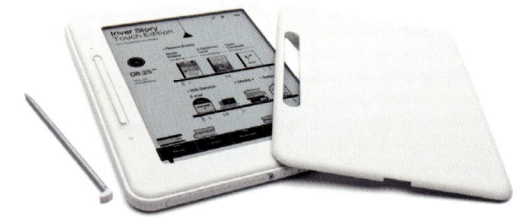

The cosmos of the i-devices: The iPad and its peripheries, filiations, and adaptations

183. iriver Cover Story, iriver, 2010.
184. Scout, Odys, 2010.
185. Reader Touch Edition, Sony, 2009.
186. OYO, Thalia booktrade group, 2010.
187. ViewSonic VEB612, ViewSonic, 2009.
188. Nook, Barnes & Noble, 2010.
189. E-Reader B630, Hanvon, 2008.
190. Kindle 3G, Amazon, 2010.

Introduced in April 2010, the iPad is a thin, flat computer, also called a »tablet«, with a touch-sensitive display measuring almost 30 cm diagonally and backlit with LEDs. This extremely lean lifestyle device, more elegant than any previous laptop or netbook, allows Internet users to receive and send e-mails, view videos, images, and TV shows, play video games and listen to music. Almost all of the approximately 200 000 programs from the app store can run on them, including those already purchased for the iPod touch or iPhone. Photographs from PCs and digital cameras can be transferred, archived, and arranged.

Above all, however, this graphic tablet aims to boost e-reading, the business of distributing electronic editions of daily newspapers and magazines. Because not only do we have the possibility for transferred print pages with static images, but also for videos linked in with the texts, linked additional information such as videotaped interviews and, of course, video commercials. Furthermore, the book or newspaper pages can be »turned« with the flick of a finger on the glass surface: digital technology thus simulates the habits of the analog world.

Apple has learned from its own mistakes and from the mistakes of other companies. In 1991 already, Sony introduced the Data Discman to the market, which used CD technology for e-reading in six to seven languages. Apple's message pad Newton (starting 1997) was an early electronic book reading device (starting mid-1994) that included a touchscreen, a stylus pen for inputing data, and handwriting recognition software. Basically, it came to market ten years too early and had such low sales numbers that Steve Jobs, after his return to Apple in 1997, stopped production. One reason for this failure was also the fact that more than one million units of the comparable Palm Pilot were sold within 18 months after its market introduction in 1996. In 2007, the online dealer Amazon launched the electronic reading device Kindle in the USA. It neither met the expectations of the publishing industry. This was first due do the relatively modest paperback size of this e-book reader, second due to the fact that books could only be scrolled vertically, and third due to the fact that the books were simply scanned in. A real media advantage compared with printed books hardly existed (the app store has been offering the application Kindle for iPhone since 2009 – a fact that does not lack irony). The iPad does not have any of these limitations: it may only be a little bit bigger, but it is multimedia compatible, and the scanned books, as mentioned before, can be leafed through horizontally or vertically. However, Amazon assiduously continues to bank on the promotion of its electronic books: at the end of August 2010 the company delivered a new version of the Kindle to stores. Recently, other manufacturers have also begun to present themselves more optimistically. Odys, for example, offers its media book Scout with a mini-display of just 13 cm; it is not only an e-book reader but can also store music, videos, and photos. America's biggest book store chain, Barnes & Noble, offers its Nook as an electronic reading device; however, to date it has not been able to efficiently compete with the Kindle and iPad. All in all, the sales numbers of e-books are currently climbing rapidly, and this is probably no coincidence following the launch of the iPad. Depending on its capacity, thousands of books can be stored on an e-book. Currently, approximately 70 000 books from a total of 1.2 million German titles are available as e-books. In the second quarter of 2010 Amazon sold more digital books than printed books for the first time. Already, it is predicted that by 2012 at the latest 25 percent of all books sold in the USA will be e-books. Sony also offers e-book readers: the Reader Pocket Edition and the Reader Touch. What remains is the question about which device people will use to read them.

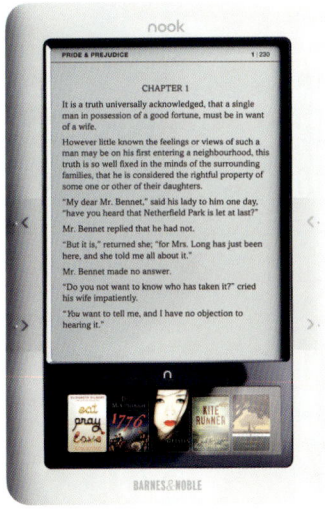

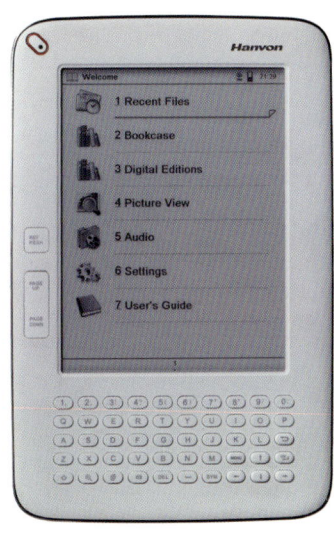

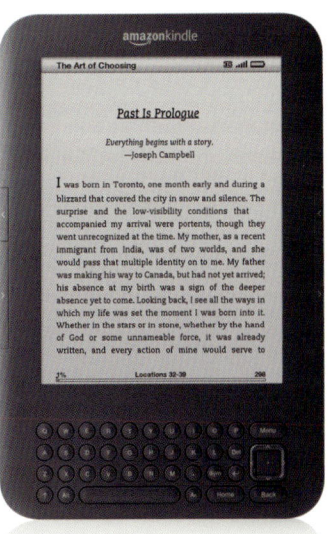

im Programm: den Reader Pocket Edition und den Reader Touch. Bleibt nur die Frage, auf welchem Gerät sie gelesen werden. Und da ist das iPad allemal attraktiver als seine Konkurrenzprodukte.

Diese iPad-Tafel im Oktavformat (24,3 x 14 cm x 1,3 cm), halb aus Glas, halb aus Aluminium, war – obwohl oder gerade weil sie eigentlich nicht mehr als ein vergrößerte Kombination zwischen iPod und iPhone, ergänzt durch ein paar Laptop-Funktionen, ist – von Anfang an ein vielleicht noch deutlicheres Objekt der Begierde als das iPhone. Durch gewissermaßen gezielte Geheimniskrämerei um den Verkaufsbeginn, die Geräte-Features und seinen Preis wurde der Hype gezielt angeheizt. Bereits am Abend vor dem ersten Tag der Auslieferung in den USA im Januar 2010 kampierten Aficionados und Nerds in Scharen vor den Apple Stores, und dies weltweit. Am ersten Wochenende nach dem Verkaufsbeginn wurden allein in den USA 500 000 Exemplare geordert.

Ulf Schönert faßt die Eigenschaften der Tablet-Computer in der Zeitschrift Stern zusammen: »Tablet-Computer sind Internet-Surfmaschinen, Fernseher, digitale Bücher, Photoalben, Spielkonsolen und Musikabspielgeräte in einem. Hunderttausende Miniprogramme (Apps), die übers Internet heruntergeladen werden können, verleihen ihnen weitere Funktionen, machen sie zu Skizzenbüchern, Videoschnittstationen, Radios, Fernsteuerungen, Zeitschriften, Shopping-Beratern, Navis und noch viel mehr. Ihrer äußeren Schlichtheit steht ein umso komplizierteres Innenleben gegenüber: Auf engstem Raum drängen sich Antennen für WLAN, GPS, UMTS, Bluetooth, Berührungssensoren, Umgebungslichtsensoren, Lagesensoren, Kameras, Mikrophone und Lautsprecher, Einschubfächer für SIM- und Speicherkarten, Graphikkarten im Miniformat, Akkus mit nie dagewesener Ladekapazität. Und nicht zuletzt Software, die jeder versteht, ohne ein Handbuch zu benutzen, und die dabei auch noch edel wirkt.«[25]

Selbstverständlich gehören auch zum iPad Peripherieprodukte, z. B. ein iPad Dock für Synchronisierungs- und Ladevorgänge, ein iPad Case als Transportschutz oder ein iPad Keyboard Dock, eine externe Tastatur für Steckdosenanschluß über ein USB-Netzteil, welches auch für Synchronisierungen mit PCs dient. Diesem digitalen Heilsbringer wurde auch in den Medien einhellig gehuldigt: Er sehe extrem elegant aus, schmücke seinen Träger, lasse sich zudem noch kinderleicht bedienen und verbinde sich jederzeit drahtlos mit dem Internet. Jordan Mejias formulierte zur amerikanischen iPad-Premiere in der Frankfurter Allgemeinen Zeitung geradezu messianisch-karthatische Bezüge. Unter der Überschrift »Die Welt leuchtet auf« heißt es weiter: »Steht die Auferstehung ins High-Tech-Paradies bevor?«[26] Und zur europäischen Markteinführung dieses Tablet-Computers im Mai 2010 veröffentlichte der Spiegel unter dem Titel »Der iKult. Wie Apple die Welt verführt« eine emblematische Analyse des Apple-Imperiums, die eine Mischung aus Hagiographie und Häme auszeichnete.[27] Das Titelbild zeigte emphatisch gereckte Hände zum Himmel hin, am oberen Bildrand blinkte der angebissene Logo-Apfel von Apple Inc. an einem Ast: eine graphisch famose Anspielung auf das Paradies und die Heilserwartungen, die die Käufer der Apple-Produkte mit diesen verbinden. Der Gründer und Chef Steve Jobs wurde in der Artikelüberschrift dann sogar hochtrabend als »Der Philosoph des 21. Jahrhunderts« gefeiert. Aber auch in diesem Artikel wurde behauptet, daß die i-Geräte sich stilistisch auf das klassische Braun-Design beziehen würden. Ich halte dies, wie gesagt, für ein Mißverständnis. Denn bei den Audiogeräten von Braun war die Gehäusegestaltung dominant. Sie zeigten geradezu idealtypisch jene »Anzeichenfunktionen«, die im Rahmen der »Produktsprachen«-Theorie analysiert wurden.[28] Bei den i-Geräten dagegen sind gerade nicht die Gehäuse, die »anzeichenfunktional« leer bleiben, sondern die interaktiven Kommunikationsketten und revolutionären Benutzeroberflächen entscheidend.

Die generalstabsmäßig geplante, zeitlich versetzte Markteinführung des iPads in Amerika und Europa löste hysterische Begehrlichkeiten, Schreikrämpfe der Nerds und nächtliche Belagerungen der Apple Stores aus. In den ersten achtzig Tagen des Verkaufs wurden weltweit drei Millionen abgesetzt und für das gesamte Jahr 2010 knapp dreizehn Millionen prognostiziert, für 2011 36,5 Millionen und für 2012 über 50 Millionen. Weitere Schätzungen besagen, daß in den nächsten Jahren weltweit Hunderte Millionen Geräte verkauft werden. Das Magazin Stern schrieb ebenso lapidar wie zutreffend: »Es ist, als wäre der PC neu erfunden worden.«[29] Fast zeitgleich zur iPad-Premiere wurden Konkurrenzprodukte angekündigt. Das Online-

191. BookBook, twelve south, 2010.
192. Data Discman, Sony, 1991.
193. Jonathan Ive: iPad, Apple, 2010.

And here, the iPad is always more attractive than competing products.

This iPad tablet in octavo format (24.3 x 14 x 1.3 cm), half glass and half aluminium – although (or especially) because it is actually nothing more than an enlarged cross between an iPod and an iPhone complemented with a few laptop functions – from the beginning was probably an even more coveted object of desire than the iPhone. The hype was intentionally fiery, and there was targeted secrecy about when it would be available, the features it would have, and its price. In January 2010, the evening before the launch of the iPad in the US, hordes of aficionados and nerds were camped outside Apple stores throughout the world. On the first weekend after the start of sales, 500 000 units were ordered in the USA alone. Ulf Schönert sums up the characteristics of the tablet computers in *Stern* magazine: »Tablet computers are all in one – Internet search engines, television sets, digital books, photo albums, games consoles, and music playback devices. Hundreds of thousands of mini-programs (apps) that can be downloaded from the Internet provide them with even more functions, and turn them into sketch books, video cutting stations, radios, remote controls, magazines, shopping guides, navigation systems, and much more. The outer simplicity of their design is countered by a highly complicated inner life: antennas for WLAN, GPS, UMTS, Bluetooth, touch sensors, ambient light sensors, position sensors, cameras, microphones and speakers, slots for SIM and storage cards, mini-format graphic cards, and unparalleled storage capacity batteries are all grouped in the smallest possible space. And, last but not least, software that anybody can understand without using a manual and that, on top of it all, looks very sophisticated.«[25]

Of course, peripheral products are also part of the iPad, for example an iPad Dock for synchronization and storage processes, an iPad Case for protection purposes, or an iPad Keyboard Dock, an external keyboard for outlet connection via a USB power adapter, which also enables synchronizations with PCs.

The media also unanimously worshipped this digital savior: they stated that it looks extremely elegant, adorns its wearer and, in addition, is child's play to operate and can be wirelessly connected to the Internet at any time. Jordan Mejias formulated almost messianic-cathartic references in *Frankfurter Allgemeine Zeitung* at the occasion of the premiere of the iPad in the US. Under the heading »The World Is Lightening Up« he writes: »Is the resurrection into high-tech paradise imminent?«[26] And *Der Spiegel* published an emblematic analysis of the Apple empire when this tablet computer was introduced in Europe in May 2010 under the title »The iCult. How Apple Seduces The World«; it stood out due to a mix of hagiography and malice.[27] The cover presented hands emphatically reaching towards the sky and, at the top of the image, the bitten into logo of Apple Inc. hanging from a branch: a graphically splendid allusion to paradise and the hopes of salvation that the buyers of the Apple products associate with them. Founder and CEO Steve Jobs was even magniloquently celebrated as »The Philosopher of the 21st Century« in the article's header. But this article also claimed that the i-devices stylistically refer to the classic Braun design by Dieter Rams. Like I said before, I consider this a misunderstanding because in the case of Braun's audio devices the design of the housing was dominant. In an almost ideally typical way, they display those »indication functions« that were analyzed within the framework of »product language« theory.[28] In the case of the i-devices, however, it is the housing that remains empty with regard to »indication functions«, and the interactive communication chains and revolutionary user interfaces are decisive. The meticulously planned, chronologically offset market launch of the iPad in America and Europe evoked hysterical envy, screaming fits from the nerds, and nightly besieges of Apple stores. During the first eighty days of sales, three million iPads were sold worldwide, with projected sales of almost thirteen million units in 2010; for 2011 this number is expected to climb to 36.5 million, and for 2012 to more than 50 million. Other estimates predict that hundreds of millions of tablet computers will be sold worldwide in the next few years. *Stern* magazine wrote: »It is as if the PC was reinvented.«[29]

Almost simultaneous with the premiere of the iPad, competing products were announced. The online company United Internet announced a smaller but also cheaper tablet computer, in a sense a »people's pad« called SmartPad, which runs Google's operating system Android and has an 18 cm monitor and USB socket. However, this cheap pad is intended to be available only as a subscription bonus in a package with its own data tariffs, meaning DSL contracts. In the beginning of August 2010 the SmartPad was already being advertised with double-page advertisements in *Focus* magazine. All of this, however, did not prevent the SmartPad from being taken off the market after only six weeks.

The company Fusion Garage offered another tablet computer, the Joojoo, but in the meantime this tablet is no longer available. The Berlin company Neofonie, rather a bonsai company on the Internet, also grandiosely announced its own pad called WeTab. Due to technical difficulties, however, the device could not be presented on three separate occasions at specially organized press conferences. In a fourth attempt, the launch was scheduled for mid-September 2010, but even at IFA 2010 in the beginning of September a functioning device could not be found. This, too, became an »i-flop«: honi soit qui mal y pense! After that, the company's managing director wrote eulogies for this dubious device on the Internet under a pseudonym, aiming to stimulate its sale, until a Berlin journalist revealed his identity. At the end of 2010, the company changed its name to 4-Tiitoo; but the software problems remained.

However, the first iPad clone, called the ePad, was produced in the Shenzhen region of China, where Apple's i-products are manufactured. Therefore, a rumor is circulating that it could just be an ironic Internet hoax. At the end of May 2010, just two or three weeks after the European market launch of the iPad, the Korean manufacturer Asus presented two tablet computers with touch-sensitive screens, the Eee Pad 101 and the Eee Pad 121. The com-

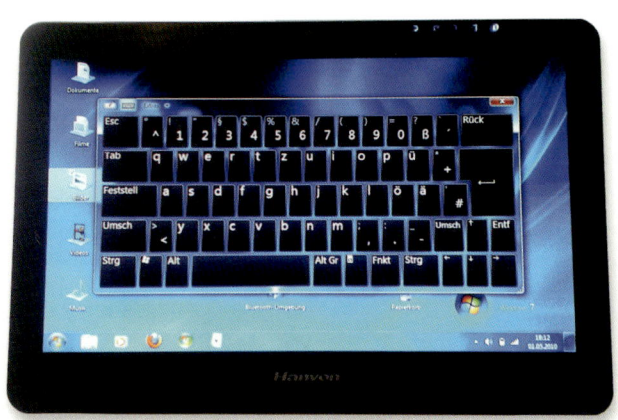

196. WeTab, 4tiitoo, 2010.
197. eee Pad 101, Asus, 2010.
198. BC10, Hanvon, 2010.
199. Galaxy Tab P1000, Samsung, 2010.
200. 101, Archos, 2010.
201. ViewSonic, View Pad, 2010.
202. Internet-Tablet Journ.E Touch, Toshiba, 2010.
203. SmartPad, 1&1 Internet AG, 2010.
204. ZiiO 10, Creative, 2010.
205. Windpad, MSI Computers, 2010.
206. Streak, Dell, 2010.

Unternehmen United Internet avisierte einen kleineren, aber auch billigeren Tablet-Computer, gewissermaßen einen »Volkspad«, genannt SmartPad, der mit dem Google-Betriebssystem Android läuft, einen 18-cm-Bildschirm und USB-Anschluß hat. Allerdings soll es dieses Billig-Pad nur als Abonnement-Prämie im Paket mit eigenen Datentarifen, also DSL-Verträgen, geben. Anfang August 2010 wurde das SmartPad bereits mit doppelseitigen Anzeigen im Magazin Fokus beworben. All dies verhinderte allerdings nicht die Peinlichkeit, daß nach gerade einmal sechs Wochen das SmartPad vom Markt genommen wurde. Das Unternehmen Fusion Garage bot einen weiteren Tablet-Computer, das Joojoo an, der aber inzwischen nicht mehr auf dem Markt ist. Auch die Berliner Firma Neofonie, im Internetgeschäft eher ein Bonsai-Unternehmen, kündigte vollmundig ein eigenes Pad namens WeTab an. Wegen technischer Schwierigkeiten konnte allerdings zwei oder dreimal bei eigens einberufenen Pressekonferenzen das Gerät nicht vorgeführt werden. Im vierten Anlauf sollte der Start Mitte September 2010 stattfinden, aber selbst auf der IFA 2010 Anfang September gab es noch kein funktionsfähiges Gerät. Aber auch das wurde eher ein »i-Flop«: Honi soit qui mal y pense! Daraufhin schrieb der Geschäftsführer dieses Unternehmens unter einem Pseudonym im Internet Elogen auf das zweifelhafte Gerät, um den Verkauf zu stimulieren, bis ihn ein Berliner Journalist enttarnte. Ende 2010 wurde die Firma in 4tiitoo umbenannt, aber die Softwareprobleme gibt es nach wie vor.

Allerdings kam der allererste, wie man auch sagt, iPad-Klon aus China, genannt ePad, typischerweise hergestellt in der Boom-Region Shenzhen, also da, wo auch Apple seine i-Produkte herstellen läßt. Deshalb kursiert die Vermutung, es könnte sich auch nur um eine ironische Simulation im Internet handeln. Ende Mai 2010, gerade einmal zwei, drei Wochen nach der europäischen Markteinführung des iPad, hatte der koreanische Hersteller Asus zwei Tablet-Computer mit Touchscreen präsentiert, den Eee Pad 101 und den EeePad 121. Die Firma Hanvon hat die iPad-Klone B 10 und B 20 im Programm, LG den Tablet-Computer UX 10 und Pioneer das DreamBook ePad A 10. Das japanische Unternehmen Sharp kündigte für den Herbst 2010 ebenfalls ein iPad-ähnliches Gerät an. Auch Acer und Toshiba (Folio 100) haben sich mit eigenen Tablet-PCs aus der Deckung gewagt. Die Firma DigitalRise hat das X9 slate tablet im Programm, ExoPC ein Tablet, das ebenfalls Slate heißt, Cisco und Netbook weitere iPad-Varianten. Mitte August brachte Dell in den USA seinen ersten Tablet-Computer, den Streak, auf den Markt. Dieses Gerät ist deutlich kleiner als das iPad, kann im Gegensatz zu diesem aber auch telephonieren. Selbst der Softwarekonzern Microsoft arbeitet an Konkurrenzmodellen zum iPad. Zur Berliner Funkausstellung 2010 stellte Samsung sein Galaxy Tab vor, welches ebenfalls mit dem Betriebssystem Android läuft und bereits zur Markteinführung über eine umfangreiche Anwendungssoftware verfügt, z. B. über die Anwendung Readers Hub mit mehr als 2 Millionen E-Books in 25 Sprachen oder einen Music Hub genannten Online-Musikkanal mit mehr als 10 Millionen Musiktiteln, der Apples iTunes Konkurrenz machen soll. Auch Hewlett-Packard, Research In Motion®, ViewSonic aus den USA mit seinem Viewpad 7, Philips mit dem Go Gear Connect, das französische Unternehmen Archos sowie Hannspree aus Taiwan haben Tablet-Computer in ihren Produktportfolios oder entwickeln sie gerade. Etwas nostalgisch kombiniert der Hersteller Cisco einen abnehmbaren Tablet-Computer mit einem Schnurtelephon in der Anmutung der 1970er Jahre. Der Spiegel titelte zum rasanten Anwachsen der iPad-Klone gewohnt süffisant: »Angriff der Trittpadfahrer.«[30] Manche dieser Fahrer, so scheint es, haben ihre Fahrlizenz wohl eher auf Verdacht. Dieser Verdacht erhärtet sich, wenn man liest, daß auf der Consumer Electronics Show (CES) in Las Vegas im Januar 2011 über fünfzig (!) neue Tablet-Computer vorgestellt wurden.

Auch die ersten »Docking Stations« für das iPad ließen nicht lange auf sich warten. Das Gerät Fidelio DS 8550 von Philips gibt Musik und Ton von Spielen oder Videos wieder und das Altec Lansing Octiv 450 ermöglicht neben dem Raumklang auch, verschiedene Neigungswinkel des iPads einzustellen und es wahlweise vertikal oder horizontal zu nutzen. Und schon kommt auch der aktuelle Trend zum Dreidimensionalen bei den Tablet-Computern zum Tragen. Die chinesische Firma Rockchip stellte bereits als Studie ein 3D-Tablet vor, welches ohne Spezialbrille auskommt und

207. Aspire 1825PT, Acer, 2010.
208. LePad, Lenovo, 2011.
209. Cisco Cius, Cisco, 2011.

pany Hanvon has the iPad clones B 10 and B 16 in its program, LG the tablet computer UX 10, and Pioneer the DreamBook ePad A 10. The Japanese company Sharp also announced an iPad-like device for the fall 2010. Acer and Toshiba (Folio 100) have come out of the gate with their own tablet PCs as well. The company DigitalRise has the X9 slate tablet in its program, ExoPC a tablet that is also called Slate, Cisco and Netbook offer more iPad versions. In mid-August 2010 Dell introduced its first tablet computer, the Streak, in the USA. This device is clearly smaller than the iPad, but it can also be used for making phone calls. Even Microsoft is working on models to compete with the iPad. At the Berlin IFA 2010 Samsung presented its Galaxy Tab, which also runs with the Android operating system and already included a comprehensive range of applications at its market launch – for example Readers Hub with more than 2 million e-books in 25 languages, or an online music channel named Music Hub with more than 10 million music titles, which aims to compete with Apple's iTunes. Hewlett-Packard, Research In Motion®, ViewSonic from the USA with its Viewpad 7, Philips with the Go Gear Connect, the French company Archos and Hannspree from Taiwan all have tablet computers in their product portfolios or are currently developing them. Almost nostalgically, the company Cisco combines a portable tablet computer with a conventional 1970s-style stationary telephone. *Der Spiegel*, in its typically smug way, headlined the rapid growth of the iPad clones: »Attack of the Copypads.«[30] Some of these »cats«, it seems, received their seven lives based on suspicion. This suspicion is confirmed by the information that more than fifty (!) new tablet computers were presented at the Consumer Electronics Show (CES) in Las Vegas in January 2011.

The first docking station for the iPad also came along rather quickly. Fidelio DS 8550 from Philips plays back the music and audio of games or videos, and the Altec Lansing Octiv 450 in addition to stereophonic sound also enables the setting of different angles of incline, allowing its vertical or horizontal use. The current trend towards three-dimensionality is already coming to bear with the tablet computers. The Chinese company Rockchip has already presented a 3D tablet as a study; it does not require special glasses and it refers to Nintendo's 3D games console in its technology. Incidentally, the iPads, similar to the iPods, instantly became popular bonus gifts for new newspaper subscribers. It is not without a certain irony that the subscribers who choose an iPad as a bonus gift also get to enjoy the electronic newspaper free of charge, while the poor regular new subscriber has to pay for at least one year of the print version.

However, iPads were not – yet – welcome at the Bundestag in mid-2010, at least not at the speaker's desk. At the time, the Procedures Committee of the German Bundestag prohibited a member of the Bundestag from using an iPad for his speech instead of a manuscript. However, by the end of the year almost 200 devices were circulating through the Bundestag, i.e., one third of all parliamentarians were using iPads.

In terms of design theory, the iPad is yet another example of convergence, i.e., the integration of previously different devices with different functions into a »uniform device«, but also an example of the »disruptive technologies« mentioned at the beginning of this text. Just as CDs displaced audio cassettes and DVDs replaced video cassettes, the iPad will displace the flat screens of TVs, PCs and laptops sooner or later: today, already, a considerable amount of television consumption already takes place via PC monitors, just like telephoning. And the »e-postal letter« is already beginning to replace paper correspondence in Germany. Phone providers in Germany such as Telekom, Vodafone and O2 have already or are about to launch their own Internet television offers. On these devices, applications from the mobile Internet worlds such as apps or social networking sites can run and be integrated in the TV offer. At the end of the year 2010 one million users had already registered, among them big companies and state authorities. The German postal service's E-Postbrief is as simple as e-mail and as secure as a letter. The encoded content can be neither read nor altered. Spams are no longer possible, and if the recipient is not registered in the postal letter portal the sender addresses him or her via the postal address. Deutsche Post then prints out the letter and delivers it in the conventional way.

sich technisch an die mobile 3D-Spielkonsole von Nintendo anlehnt.

Im übrigen sind ähnlich wie die iPods auch die iPads sofort zu beliebten Prämien für die Neuwerbung von Zeitungsabonnenten geworden. Dabei entbehrt es nicht einer gewissen Ironie, daß der Werber, der sich als Prämie ein iPad aussucht, damit gratis in den Genuß auch jener elektronischen Zeitung kommt, die der arme Neuabonnent als Printversion mindestens ein Jahr lang bezahlen muß. Im Bundestag allerdings waren iPads, zumindest am Rednerpult, noch Mitte 2010 nicht willkommen. Damals untersagte der Geschäftsordnungsausschuss des Bundestages einem Abgeordneten, der statt eines Manuskripts ein iPad für seine Rede benutzen wollte, dessen entsprechende Benutzung. Am Jahresende aber waren bereits ca. 200 Geräte im Umlauf, d. h. ein Drittel aller Abgeordneten nutzt das iPad.

Designtheoretisch ist das iPad ein weiteres Beispiel für die Konvergenz, also für das Integrieren vormals verschiedener Geräte mit differenten Funktionen in einem »Einheitsgerät«, aber ebenso ein Beispiel jener am Beginn dieses Textes benannten »disruptiven Technologien«. So wie die CD die Audiokassette verdrängt hat, die DVD die Videokassetten, so wird über kurz oder lang das iPad die Flachbildschirme der TVs, PCs und Laptops verdrängen: Schon heute findet ja ein erklecklicher Teil des Fernsehkonsums auf den PC-Monitoren statt wie übrigens auch das Telephonieren. Und schon schickt sich der E-Postbrief an, die Papierkorrespondenz zu ersetzen. Telephonanbieter in Deutschland wie Telekom, Vodafone oder O2 haben bzw. starten bald eigene Angebote für Internetfernsehen. Auf solchen Geräten können auch Anwendungen aus der mobilen Internetwelt wie Apps oder Communities laufen und mit dem TV-Angebot verbunden werden. Am Ende des Jahres 2010 hatten sich bereits eine Million Nutzer angemeldet, darunter auch Großunternehmen und Behörden. Der E-Postbrief ist so einfach wie eine E-Mail und so sicher wie ein Brief. Die verschlüsselten Inhalte kann man weder mitlesen noch verändern. Spams sind nicht mehr möglich und wenn der Empfänger im Postbrief-Portal nicht registriert ist, adressiert der Absender ihn mit der Postanschrift. Die Deutsche Post druckt den Brief dann aus und stellt ihn konventionell zu.

Zur Apple-internen Konvergenz gehört auch, daß das Unternehmen nach den Erfolgen von iPhone und iPad über berührungsempfindliche Bildschirme für seine Tischcomputer nachdenkt. Solche Touchscreen-Macs müßten allerdings kippbare Bildschirme aufweisen, um bequem mit den Fingern bedient werden zu können. Vorsorglich hat Apple auch dafür bereits ein Patent angemeldet. Inzwischen hat Apple auch den erfolgreichen App Store von iPhone und iPad auf seine PCs und Notebooks übertragen. Die Software-Vertriebsplattform Mac App Store ist damit Teil des Betriebssystems OS X geworden.

Im übrigen hatte frog design bereits 1983 einen »Prä-Touch-Tablet-Computer für Apple entwickelt. Er hieß nach einem der Zwerge des Märchens *Schneewittchen* Bashful und bezog sich so schon durch den Namen auf die von frog design Snow White genannte Designsprache, die sie für die Serie der Apple Ii-Computer entwickelt hatten. Ein Konzept für dieses Tablet in Keilform – eine Entwurfskonzept, welches an die zeitgleichen Schreibmaschinen von Mario Bellini für Olivetti erinnert – sah eine externe Tastatur vor, ein anderes Konzept integrierte ein Diskettenlaufwerk. Ein dazu gehörender Stift sollte die Interaktion des Nutzers mit dem Bildschirm erleichtern. Ein weiteres Konzept integrierte ein Telephon. In Hinsicht auf diese frühen Entwürfe bemerkte Hartmut Esslinger, der Gründer von frog design und Chefentwerfer der Snow White-Geräte, leicht ironisch: »Yes, good innovation can take a while.« Dieser kurze Blick ins Archiv der frog-Designer erschließt gewissermaßen einen Teil der Archäologie der Tablet-Computer. In den Tagen jener frühen Versuche war kaum vorherzusehen, daß inzwischen pro Tag 230 000 neue iPhones und iPads verkauft werden. Gegenwärtig sind 120 Millionen dieser Geräte in Gebrauch.

210. Fidelio DS9000, Philips 2010.
212. Hartmut Esslinger: prototype Big Apple TouchPad Computer, 1984.
211. Octiv 450, Altec Lansing, 2010.
213, 214. Hartmut Esslinger: prototype Apple Bashful Tablet Mac, 1983.
215. Hartmut Esslinger: prototype MacPhone – NotePad 1, 1984.

Also part of Apple's internal convergence is the fact that the company, after the phenomenal success of the iPhone and iPad, is considering offering touch-sensitive monitors for its desktop computers. However, such touchscreen Macs have to provide flip-over monitors in order to be easily operated with the fingers. Apple has already registered a patent as a precaution. In the meantime, Apple has transferred the successful iPhone and iPad App store to its PCs and notebooks. The software sales platform Mac App store has thus become part of the operating system OS X.

Incidentally, frog design already developed a pre-touch tablet computer for Apple in 1983. It was named after one of the dwarfs in the fairy tale *Snow White*: Bashful thus already referred to the design language called Snow White by frog design, which they had developed for the Apple II computer series. One concept for this wedge-shaped tablet – a design concept that is reminiscent of the typewriters by Mario Bellini for Olivetti from the same time – required an external keyboard; another concept integrated a disk drive. A pen was to facilitate the interaction of the user with the monitor. Another concept had an integrated telephone. With regard to these early designs, Hartmut Esslinger, the founder of frog design and chief designer of the Snow White devices, remarked with a touch of irony: »Yes, good innovation can take a while.« This short glance into the archives of the frog designers in a sense taps into part of the tablet computers' archaeology. In the era of those early efforts it was almost impossible to predict that there would come a time when 230 000 new iPhones and iPads per day would be sold. Currently, 120 million of these devices are in use.

Der i-Kosmos: Das Markenimage, das Marketing, die i-Apps und die Apple Stores

Wie bei allen großen, international tätigen Unternehmen gibt es auch bei Apple Kritiker, die ein paar dunkle Flecken auf der blütenweißen (in diesem Fall wohl eher Snow White-) Weste der Firma ausgemacht haben wollen. So betreibe das Unternehmen eine recht rigide Copyright-Politik, denn sowohl beim iPhone wie beim iPad können nur von Apple genehmigte Anwendungen benutzt werden. Apple läßt seine i-Produkte bei der taiwanesischen Firma Foxconn in Shenzen nahe Hongkong fertigen, die als größte Elektronikfabrik der Welt mit 900 000 Arbeitern auch als iPod-City bezeichnet wird. Man kann sagen, daß die Stückkosten bei einem »Apple und 'nem Ei« liegen. Ein Arbeiter verdient bei Zwölfstundenschichten etwa 200 Euro im Monat, also weniger, als ein iPhone kostet. Nach einigen Selbstmorden von Foxconn-Mitarbeitern reklamierte Apple besorgt moralische Mindeststandards in dem Unternehmen, stellte auch eigene Untersuchungen an und verwies auf seinen eigenen »Supplier Code of Conduct«.[31]

Auch die interne Unternehmenskultur, so wird berichtet, sei von einer eher strengen Mentalität geprägt. So wie viele die Apple-Produkte eigentlich nicht brauchen, aber begehren, »also doch brauchen. Brauchen wollen.«[32] so haben die Mitarbeiter der Firma Angst vor dem Entzug von Wärme, Angst vor Mißachtung durch den Chef Steve Jobs: »Daß er dich anhört, dir ›face time‹ gibt, ist Zeichen der Bedeutung deiner Aufgabe. Die iTunes-Store-Leute, jene, die den interaktiven Laden für Musik und Filme aufbauten, bekamen vor rund fünf Jahren eine Menge ›face time‹. Im Moment sind die iPad-Jungs und -Mädchen modern.«[33] Die Kleiderordnung scheint wesentlich liberaler als die Mitarbeiterhierarchien zu sein. Aber dies sind nur Facetten eines Firmenimages, welches schlußendlich auf der Faszination seiner Produkte beruht; -Facetten zudem, die die interne Firmenkultur, die generell von Respekt und Wärme gekennzeichnet ist, kaum relativieren können.

Gewissermaßen an den entgegengesetzten Endpunkten zwischen den analogen und den digitalen Welten von Apple steht die Hardware der Firmenarchitektur und die Produktwerbung einerseits und die Software der Anwendungsprogramme, der Apps, andererseits. Diese sind zwar herstellergebunden und können nicht beliebig auf alle Smartphones installiert werden, oft aber auf mehreren. Außer Apple bieten andere Hersteller wie Research In Motion® (RIM®) für den BlackBerry® oder Google für sein Android-System jeweils eigene Apps an.

Neben dem Firmensitz in Cupertino in Kalifornien, einer messebauartigen Glashalle mit gewölbtem Tonnendach sind die Apple Stores die Hauptträger der architektonischen Corporate Identity des Unternehmens. Es sind oft vollständig gläserne Hüllen, frei stehend auf Plätzen, wie z. B. in New York ein Glaswürfel oder in Shanghai eine Glastonne, die im Zentrum eines abgesenkten kreisrunden Platzes mitten im Boomviertel Pudong steht, dessen umlaufende Begrenzung wie ein Amphitheater angelegt ist. Gläserne Freitreppen führen jeweils in die unterirdischen Verkaufsräume. Die bis auf die abgehängten, leuchtenden Apfel-Logos praktisch leeren Glashüllen verweisen so auf Transparenz, Immaterialität und Virtualität. Aber eben auch auf Exklusivität, edle Gestaltung und selbstbewußte Gediegenheit: im Sinne des Begriffs einer logo architecture. Dieser transparente Minimalismus ent-

216. Apple Store Shanghai, Pudong, 2010.
217. Apple Store München, 2008.
218. Apple Store Beijing, 2010.
219. Apple Store New York, 2006.
220. Apple Store Convent Garden, London, 2010.
221. Apple Store Opéra, Paris, 2010.
222. Apple Store Sydney, 2008.

The i-cosmos: The brand image, the marketing, the i-apps and the Apple stores

As with all big, internationally active companies, Apple also has critics who claim to have found a few dark spots on the clean (in this case rather Snow White) slate of the company. They say that it is running rather rigid copyright policies because only applications approved by Apple can be used for the iPhone and the iPad. Apple has its products manufactured by the company Foxconn in Shenzhen near Hong Kong; with 900 000 workers, it is the world's biggest electronics factory and also called iPod City. One could say that the production cost of an Apple product is very reasonable. Working twelve-hour shifts, workers earn approximately 200 euro per month, which is less than the cost of an iPhone. After several suicides of Foxconn workers, Apple anxiously demanded minimum moral standards at the company, undertook its own investigations, and pointed out its own »Supplier Code of Conduct.«[31]

As is reported, the internal corporate culture is also characterized by a rather strict mentality. Just as many do not really need but desire the Apple products, »hence need them after all. Want to need them«,[32] the employees of the company are afraid of the withdrawal of warmth, of disregard by their boss Steve Jobs: »If he listens to you, gives you ›face time‹, it is a sign of the importance of your job. The iTunes store people, those who developed the interactive store for music and films, got a lot of ›face time‹ around five years ago. At the moment, the iPad guys and girls are *en vogue*.«[33] The dress code seems to be much more liberal than the employee hierarchies. But these are only facets of a corporate image that in the end is based on the appeal of its products; in addition, they are facets that can hardly make the internal corporate culture, which is generally characterized by respect and warmth, more relative.

The hardware of the corporate architecture and product advertising on the one hand and the software of the application programs, the apps, on the other in a sense stand at opposite poles of Apple's analog and digital worlds. They may be bound to the manufacturer and cannot be installed at will on all smartphones, but often on several of them. In addition to Apple, other manufacturers such as Research In Motion® (RIM®) offer their own apps for the BlackBerry® or for Google's Android system.

Alongside the corporate headquarters in Cupertino, California, a trade fair design-like glass hall with a vaulted barrel roof, the Apple stores are the main carriers of the company's architectural corporate identity. Often they are freestanding glass shells situated in squares such as the glass cube in New York or a glass barrel in Shanghai, which is situated in the center of a sunken circular square in the midst of the booming quarter of Pudong, the encircling limitation of which is designed like an amphitheater. Glazed flights of stairs lead into the subterranean sales rooms. The practically empty glass shells – except for the suspended glowing apple logos – thus refer not only to transparency, immateriality, and virtuality but also to exclusivity, noble design, and self-confident sophistication: in the sense of the term of a logo architecture. This transparent minimalism congenially complies with the fine, coolly elegant reductionism of Apple's product worlds, but also visualizes the immateriality of the apps and abstract touch displays in an equally perfect way. The Apple store in New York is open 24 hours a day, 365 days a year. Naturally,

spricht kongenial dem vornehmen, cool-eleganten Reduktionismus der Produktwelten von Apple, aber visualisiert ebenso perfekt die Immaterialität der Apps und der abstrakten Touch-Displays. Der Apple Store in New York hat 24 Stunden am Tag und 365 Tage im Jahr geöffnet. Selbstredend hat das Personal sich, wenn auch casual, einheitlich zu kleiden und ebenso selbstverständlich haben alle Präsentationstische ein kohärentes Erscheinungsbild. Die strenge Nüchternheit der kubischen, hellen Naturholztische betont umso mehr die Eleganz der offen auf ihnen präsentierten Apple-Produkte. Dieses Displaykonzept ist weltweit in allen Apple Stores gleich, ob in Shanghai oder Tokyo, in New York oder Zürich, in München, Hamburg oder Frankfurt. Der größte Store bisher befindet sich in der Regent Street in London. Es gibt weltweit über 280 Apple Stores in über 25 Ländern, vor allem in den USA, Kanada, China, Japan, Großbritannien, Deutschland, Italien, Frankreich, der Schweiz und Australien. Im Jahr 2008 erwirtschaftete Apple mit seinen Stores einen Gewinn von 920 Millionen Euro.

Mindestens ebenso wie seine Corporate Identity zeigt die Werbung eines Unternehmens für sich und seine Produkte sein Selbstverständnis. Beispielhaft werden im Folgenden einige Print- und TV-Werbekampagnen vorgestellt. Der Fokus des Interesses liegt dabei auf den sich wandelnden verbalen und visuellen Rhetoriken. Die Produktwerbung von Apple, ob als gedruckte Anzeige oder TV-Spot, war von Anfang an frech und mußte es auch sein. In der Mitte der 1970er Jahre galt es, den übermächtigen Konkurrenten IBM – der nicht umsonst »Big Blue« hieß – Paroli zu bieten. Nicht nur der Apple-Slogan »Think different« reagierte auf den IBM-Slogan »Think«, nicht nur das Apple-Regenbogenlogo paraphrasierte ein IBM-Logo, sondern auch manche Werbeanzeigen kokettierten mit direkten oder indirekten Verweisen auf IBM. So wurde z. B. in einer Marketingbroschüre von 1983 formuliert: »1977 kam der Apple II. 1981 kam der IBM PC. ... Im Folgenden werden wir fünf der häufigsten Computerfunktionen vergleichen, ... um Ihnen den Unterschied zwischen einem IBM PC und einem Macintosh zu verdeutlichen.« Neben der Textargumentation sind jeweils die Bildschirme der beiden PCs mit den fünf angesprochenen Funktionen abgebildet. Bei der Funktion »Textverarbeitung« zeigt der IBM-Schirm kryptische Programmierer-Befehlsketten, der Apple-Bildschirm die Fließtexteinführung für das Programm Mac Write. Bei der Funktion »Graphik« zeigt der Apple-Bildschirm Microsoft-Icons, der IBM-Schirm bleibt dunkel mit der sarkastischen Bildunterschrift: »Business graphics before Macintosh«.

Eine solche vergleichende Werbung wurde in Europa lange abgelehnt und stößt bis heute auf marktrechtliche Bedenken. Apple hatte diese Bedenken nicht. Da heißt es weiter: »Wenn es um Business-Graphik geht, muß man bei aller Fairneß sagen, daß IBM nur leere Versprechungen und gesperrte Charts anbietet. ... Wenn Sie die Computer vergleichen, werden Sie feststellen, daß der IBM PC nicht nur leere Versprechungen liefert, sondern daß dessen Display vollständig leer bleibt.« Am Ende der Kampagne liest man: »Es gibt nur eine Sache, die der IBM PC genauso gut kann wie der Macintosh: Er eifert 3278 Terminal-Funktionen nach – so können Sie immerhin mit den stämmigeren IBMs kommunizieren.« »Big Blue« war und blieb lange Zeit der große Gegner.

Einer der berühmtesten TV-Spots aller Zeiten war 1984 zur Markteinführung des Apple Lisa eine weitere Attacke auf IBM, die sich auf George Orwells Roman *1984* bezog. Die Idee für diesen Werbefilm stammte von dem Art Director Brent Thomas und dem Texter Steve Hayden, beide von der Werbeagentur Chiat-Day, die die ersten TV-Commercials für Apple kreierte. Der Hollywood-Regisseur Ridley Scott, der den berühmten Film *Blade Runner* (1982) machte, filmte diesen Werbespot. Zunächst sieht man in extremen Nahaufnahmen Kolonnen von marschierenden Zivilisten, die von martialisch wirkenden Soldaten angetrieben werden. Sie werden in einen übergroßen Versammlungsraum genötigt, um sich über einen Großbildschirm indoktrinieren zu lassen. Nun stürmt eine schnell laufende, leicht bekleidete Blondine, mit einem langstieligen Hammer bewaffnet, verfolgt von finsteren Polizisten, diese Versammlungshalle, in der die roboterhaft aufgereihten kahlgeschorenen Untertanen in Sträflingskleidung zwanghaft auf den Bildschirm starren, auf der der »Große Vorsitzende« Verhaltensregeln statuiert. Bis auf die blonde Frau ist die gesamte Szenerie »grau in grau« gehalten. Schon dies ist ein subtiler Hinweis auf die grauen IBM-Rechner. Mit einer Hammerwerfer-Drehbewegung schleudert die Blondine ihr Gerät wie eine Waffe auf den Großbildschirm, der krachend zerbirst. Wie mit Rauhreif überzogen, wird das Szenario noch blasser als vorher, die Sitzenden wirken ausgebleicht wie Olme. Am Ende des Spots sieht und hört man als Schrift und Ton: »On January 24th, Apple Computer will introduce Macinthosh. And you'll see why 1984 won't be like *1984*.« Symbolisch gelesen, hat damit David den Goliath besiegt. Zen-Buddhismus: Das Kleine bleibt nicht klein und das Große nicht groß. Alle Zuschauer verstanden diese Botschaft. Nicht alle jedoch sahen die Parallele zu Stanley Kubricks Film *2001 – Odyssee im Weltraum* (1965–1968). In diesem Film schleudert ein Neandertaler einen Knochen in die Luft, der bis in den Weltraum fliegt und sich dort in ein Raumschiff verwandelt. Die visuellen Parallelen liegen auf der Hand: Strahlt im Kubrick-Film das Archaische ins Hyperindustrielle und wird diesem als Erinnerung eingeschrieben, so reklamiert im TV-Spot der Newcomer eine technogenetische und kommunikative Überlegenheit über den bisherigen Platzhirsch. Es ist so, als würde Apple sagen: Wir haben die bessere DNS!

Gegenüber den frühen Werbeattacken von dem damals noch winzigen Unternehmen Apple gegen IBM sind natürlich die heutigen Print-Kampagnen meilenweit entfernt und wesentlich gelassener. Ab etwa 1990 war die textlastige Argumentation, die immer einen Rechtfertigungscharakter hat, so gut wie bei allen Unternehmen nicht mehr gefragt. Das Vertrauen auf eine dominant visuelle Rhetorik mag dabei nicht zuletzt just den graphischen Bedienoberflächen von Apple-Computern geschuldet sein. Insofern ist es auch evident, daß Apple die Marketingstrategien generell verändert hat.

Zwar wird der iPod durchaus noch mit einer sachlichen, fast lexikalischen Anzeige im Internet be-

the staff has to dress – although casually – in a uniform way, and just as naturally, all presentation tables have a coherent appearance. The strict sobriety of the cubic, light natural wood tables even more emphasizes the elegance of the Apple products that are openly presented on them. This display concept is identical in all Apple stores worldwide, whether in Shanghai or Tokyo, New York or Zurich, Munich, Hamburg or Frankfurt. The biggest store to date is located in London's Regent Street. There are more than 280 Apple stores in over 25 countries worldwide, mainly in the USA, Canada, China, Japan, Great Britain, Germany, Italy, France, Switzerland, and Australia. In the year 2008, Apple generated profits of 920 million euro through these stores.

The advertising of a company of itself and its products, at least as much as its corporate identity, presents its self-image. Below, some print and TV advertising campaigns will be described as examples. The focus of interest is on the changing verbal and visual rhetoric strategies. Apple's product advertising, whether as print ads or TV spots, was fresh from the beginning, and it had to be. In the mid-1970s the task was to vie with the overpowering competitor IBM, which was called »Big Blue« for a reason. Not only was Apple's slogan »Think different« a reaction to IBM's slogan »Think«, and not only did the Apple rainbow logo paraphrase an IBM logo, but some print ads also played with direct or indirect references to IBM. For example, a marketing brochure from 1983 stated: »In 1977 we had the Apple II. In 1981 the IBM PC. … We're going to compare five of the most typical functions a computer performs … to show you the difference between … an IBM PC and … a Macintosh.« Next to the text argumentation the five screens of the two PCs with the five functions are shown. For the function »word processing« the IBM screen shows cryptic programming command chains, the Apple screen shows the running text introduction for the program Mac Write. For the »graphics« function the Apple screen shows Microsoft icons, the IBM screen is dark and carries the sarcastic caption: »Business graphics before Macintosh.« Such comparative advertising has long been rejected in Europe and is still subject to market rights restrictions. Apple did not have these doubts. It continues: »When it comes to business graphics, in all fairness, IBM has pic and bar charts to spare. … When you compare the actual unit you purchase initially with your Macintosh, the IBM PC not only comes up short a few pie charts, it draws a complete blank.« The campaign concludes: »There is one thing that the IBM PC manages to do as well as a Macintosh: IBM 3278 terminal emulation – so you can communicate with heftier IBMs.« For a long time, »Big Blue« remained the big enemy. One of the most famous TV spots of all time was another attack on IBM at the occasion of the market launch of the Apple Lisa; it referred to George Orwell's novel *1984*. The idea for this advert came from art director Brent Thomas and ad writer Steve Hayden, both from the advertising agency Chiat-Day, which created the first TV commercials for Apple. Hollywood director and producer Ridley Scott, who directed the famous movie *Blade Runner* in 1982, filmed the spot. First, extreme close-ups show queues of marching civilians being driven onward by soldiers. They are forced into an oversized meeting room to be indoctrinated via a large screen. Then, a scantily clad blonde woman armed with a long-handled hammer, followed by somber policemen, storms into the meeting hall, where the robot-like lined up and shaven subjects compulsively stare into the screen, watching the »great chairman« laying down rules of behavior. Except for the blonde woman, the entire scene is in gray tones. This already subtlely hints at the gray IBM computers. With the moves of an Olympic hammer thrower the blonde woman throws her hammer into the big screen, which shatters into a million pieces. Looking as if covered with hoarfrost, the entire scenario turns even paler than before, and the people sitting look

223. Brent Thomas, Steve Hayden: TV-Spot for Apple II PC Lisa, Apple, 1984, produced by Ridley Scott.
224. Advertisement Macintosh SE, Apple, 1987.

worben, die alle Varianten des Gerätes plus dem iPhone zeigen, so daß ein unmittelbarer Größenvergleich möglich ist. Andererseits werden unter der bunten Überschrift »nano-chromatic« neun iPod nanos in den Farben des Regenbogens aufgereiht, bei denen die entsprechenden Farben als Lackschlieren aus den Geräten tropfen. Damit werden emotionale Assoziationen angesprochen und es wird auf eine Subkultur der Graffitis angespielt, die solche Sachbeschädigungen eher für eine läßliche Sünde halten. Mit einem solchen »Wunschfarben-Konzert« werden die Geräte zudem in den Kontext der modernen Malerei von Jackson Pollock bis Sandro Chia gestellt: eine kulturelle Nobilitierung. Schließlich betont die Chromatik die Individualität der Geräte.

Die Werbung für das iPhone hat ein logisches Dilemma zu meistern: Einerseits soll das Gerät als Mobiltelephon gezeigt werden, also am Ohr; andererseits sollen auch die Fingerspreizungen auf dem Display sichtbar werden. Die Werbung reagiert salomonisch, indem sie das Sprechen auf Freisprechen umpolt: So kann das Gerät beim Sprechen auf dem Tisch liegen. Der deutsche Kabarettist Karl Valentin würde fragen: »Ist nicht auch das Ohr eine Oberfläche, also ein Display?« oder vielleicht auch, auf die Touch-Oberflächen bezogen: »Früher war das Streicheln ganz einfach besser.«

Überhaupt ist die Printwerbung für interaktive Geräte immer insofern problematisch, weil die Fähigkeiten der Geräte allenfalls beschrieben, aber kaum visualisiert werden können. In einer merkwürdigen Volte entzieht sich die potenzierte Benutzbarkeit, das »Handling« einer adäquaten Beschreibung. In solchen Fällen produziert, ja provoziert Pragmatik Abstraktion. Die i-Geräte belegen diesen Vorgang: Sie sind Konkretions- und Abstraktionsmaschinen in einem, insofern sie die gesamte Welt herbeirufen, aber eben nur über »Fingerbefehle«, Apps oder Social Networks. Dies aber läßt sich als Druckwerbung kaum vermitteln.

Die TV-Werbung für das iPad zeigt natürlich agile Nutzer, die dieses Tablet-Computer irgendwo im Freien bedienen. Da wirbeln die Finger mit den Apps um die Wette. So wird nachdrücklich vermittelt, daß man buchstäblich alles auf das Display herbeirufen kann. »Auf den Schirm« hieß der entsprechende Befehl des Kapitän Kirk in der TV-Serie *Raumschiff Enterprise*. In der Struktur und Rhetorik dieses Spots zeigt sich die generelle Entwicklungstendenz elektronischer Geräte: Weniger besitzen als nutzen, weniger speichern als jeweils aktuell abrufen. Das »in-der-Welt-sein« wird mit dem iPad dominant in den virtuellen Datenraum verlagert. Die Wirklichkeit ist multidimensional und relativ, hinter ihr liegen weitere Wirklichkeiten. Das haben wir zunächst fiktional in Filmen wie *Blade Runner* oder *Matrix* realisiert, die die Prinzipien der filmischen Narration – daß wir glauben sollen, was wir sehen, weil wir es sehen – außer Kraft setzen. Aber nun werden wir mit Geräten konfrontiert, die dies in Pragmatik, d.h. in eigene Handlungsmöglichkeiten umsetzen. Manchmal gibt es Commercials, die tiefere Wahrheiten transportieren als ihren Createuren bewußt sein mag. Die aktuelle Form der Werbung bezieht sich natürlich auf die zahllosen Apps, für die offenbar die Geräte nur noch Behälter sind.

bleached like grotto olms. At the end of the advert are the words »On January 24th, Apple Computer will introduce Macintosh. And you'll see why 1984 won't be like *1984*« in script and audio. Read symbolically, David has thus defeated Goliath. Zen Buddhism: the small does not remain small and the big not big. All viewers understood this message. However, not all of them detected the parallels with Stanley Kubrick's film *2001 – A Space Odyssey* (1965–1968). At the beginning of the movie, a Neanderthal heaves a bone into the air; it flies up into space and transforms into a space station. The visual parallels are obvious: while the archaic jets into the hyper-industrial in Kubrick's film and is inscribed in it as a memory, the newcomer in the TV spot claims a techno-genetic and communicative superiority over the former top dog. It is as though Apple were saying: we've got the better DNA!

Naturally, today's print campaigns are miles from the early advertising attacks by the formerly tiny company Apple against IBM, and they are much more relaxed. Starting in 1990, the argumentation dominated by text, which always has a justification character, was outdated at almost all companies. The trust in a predominantly visual presentation may just be owed to the graphic user interface of Apple computers. In this regard, it is also evident that Apple generally changed the marketing strategies.

The iPod may still be advertised with a factual, almost lexical ad on the Internet, which presents all versions of the device plus the iPhone, allowing an immediate comparison of sizes; however, nine iPod nanos are lined up in rainbow colors under the colorful heading »nano-chromatic«, with the appropriate colors dripping off the devices as streaks of lacquer. This appeals to emotional associations and alludes to a subculture of graffiti, which rather considers such damage to property a venial sin. With such a »color request concert« the devices are also placed in the context of modern painting, from Jackson Pollock to Sandro Chia: a cultural ennoblement. After all, the chromatics emphasize the devices' individuality.

The advertising for the iPhone has had to master a logical dilemma: on the one hand, the device has to be presented as a mobile phone, i.e., on the ear; on the other, the finger spreads on the display have to be made visible. The advertising reacts in a Solomon-like way by changing normal speaking to hands-free speaking, allowing the device to be placed on a table in the middle of a call. Karl Valentin, the German comedian, would ask: »Is not the ear also a surface, hence a display?« Or perhaps, referring to the touch-sensitive surfaces: »Caressing simply used to be better in the old days.«

At any rate, designing print advertisements for interactive devices is always a problem insofar as they can at best describe a device's abilities but are hardly able to visualize them. In a strange volte, the potentiated usability, the »handling«, withdraws from adequate description. In such cases, pragmatics produces and even provokes abstraction. The i-devices verify this process: they are machines of conciseness and abstraction in one in that they summon the entire world, but only via »finger commands«, apps, or social networks. In print advertising, however, this is difficult to communicate.

Of course, the TV ads for the iPad show agile users using the computer tablet somewhere outdoors. The fingers are whirling in a contest with the apps. This emphatically communicates that literally anything can be summoned on the display. »On screen« is one of Captain Kirk's commands in the TV series *Star Trek*. In the structure and rhetoric of this spot the general development trend of electronic devices becomes evident: own less and use more, store less and topically download more. »Being in the world« is dominantly shifted into the virtual data space with the iPad. Reality is multidimensional and relative; behind it other realities are hidden. We first fictionally realized this in films such as *Blade Runner* or *Matrix*, which invalidate the principles of cinematic narration – that we are supposed to believe what we see because we see it. Now, we are confronted with devices that transpose this in pragmatics, i.e., into their own lines of action. Sometimes, commercials transport truths that are deeper than their creators may be aware of. Naturally, the current form of advertising relates to the countless apps for which the devices are obviously only vehicles.

Speaking of apps: in early August 2010 the German electronics chain store Media Markt had a newspaper supplement that accidentally turned the company's slogan »I am not stupid« into the exact opposite: a Samsung smartphone was advertised with a picture of the device whose apps had Russian subtitles. Otherwise, of course, the text was in German. Is this just sloppiness on the part of the advertising agency or a discrete hint at the takeover of the Korean manufacturer by the Russians?

In 2001, Apple entered the in-house software programs segment for mobile devices with the music program iTunes. This entertainment software, which only works on Apple's mobile devices, allows users to download music files from the iTunes store, which can then be played on through a computer, burnt onto a CD or downloaded to an iPod or iPhone. In the meantime the iTunes store is the most popular online entertainment store in the world. It offers more than 8.5 million songs, albums, audio books, TV series and iPod games. At the end of 2010, Apple changed the motifs on the iTunes prepaid cards. Since thirteen original studio albums by the Beatles can now be downloaded from iTunes, the new cards now depict nostalgic black-and-white photos of the Fab Four.

Recently, a new function of iTunes called Genius has allowed users to establish »taste profiles« from the downloaded music files in order to enable target group specific advertising. The app Shazzan can recognize any piece of music based on a few notes of it, show the title and artist on the display, and automatically download it from the Internet. Competing with iTunes, Microsoft is running its online music shop Zune in the USA, and soon in twenty more countries – among them France, Italy, Spain, Great Britain, and Germany. Users can listen to an unlimited quantity of music from the

225. App Store, Apple, 2011.
226. App Store, Apple, 2011.
227. MobileMe, Apple, 2008.

Apropos Apps: Anfang August 2010 gab es als Zeitungsbeilage einen Werbeprospekt des Media Marktes, der unfreiwillig den Firmenslogan »Ich bin doch nicht blöd« ins Gegenteil verkehrte. Denn ein Smartphone von Samsung wurde mit einer Geräteabbildung beworben, deren Apps russisch untertitelt waren. Ansonsten war der gesamte Prospekt natürlich in deutscher Sprache betextet. Ist dies lediglich ein Indiz für die Schlampigkeit einer Werbeagentur oder doch der dezente Hinweis auf die Übernahme des koreanischen Herstellers durch die Russen?

Mit dem Musikprogramm iTunes stieg Apple 2001 in den Bereich der eigenen Software-Programme für Mobilgeräte ein. Mit dieser Unterhaltungssoftware, die exklusiv nur auf Geräten von Apple funktioniert, lassen sich Musikdateien direkt aus dem Internet laden und können dort abgespielt, auf CD gebrannt oder auf den iPod übertragen werden. Inzwischen ist iTunes der populärste Entertainment Online Store weltweit. Er bietet über 8,5 Millionen Songs, Alben, Hörbücher, TV-Serien und iPod-Spiele an. Ende 2010 stellte Apple die Motive auf den iTunes-Prepaid-Karten um. Da man sich nun dreizehn originale Studienalben der Beatles von iTunes laden kann, zeigen die neuen Karten nun nostalgische Schwarzweißphotos der frühen Fab Four.

Seit kurzem erlaubt eine neue Funktion von iTunes, Genius genannt, ein Geschmacksprofil aus den heruntergeladenen Musikdateien zu erstellen, um zielgruppenpräzise Werbung zu ermöglichen. Die App Shazzan ist in der Lage, nur aufgrund von ein paar Tönen jedes Musikstück zu erkennen, Titel und Interpret auf dem Display anzuzeigen und den Titel automatisch herunterzuladen.

Als Konkurrenz zu iTunes betreibt Microsoft seinen Online-Musikladen Zune in den USA, demnächst auch in zwanzig weiteren Ländern, u.a. in Frankreich, Italien, Spanien, Großbritannien und Deutschland. Für 10 Euro Monatsgebühr können Nutzer unbegrenzt Musik aus dem Katalog der Zune-Plattform hören, die acht Millionen Songs umfaßt. In den USA gibt es auch ein dem iPod vergleichbares Zune-Abspielgerät. Auch Amazon und die Deutsche Telekom bieten Online-Musik an. Kostenlos sind z. B. auch selbst die Apps »Geldanlage« für Tages- und Festgeld, »Baufinanzen« mit Übersichten entsprechender Banken und Finanzmakler, während die App »Geldautomaten in der Nähe« 1.59 Euro kostet, was sich allerdings angesichts gebührenpflichtigen Abhebens bei Fremdbanken allemal lohnt. Auch werden die Bankenstandorte räumlich angezeigt.

Auch für das iPad liefert ausschließlich und allein Apple die Software für Musik, Bücher und Videos. So hält man Konkurrenten in Schach und macht gute Geschäfte, zumal dann, wenn diese Software kostenpflichtig ist. Zu dieser rasant wachsenden Zahl der kostenpflichtigen Apps zählt Apples Online-Dienst MobileMe, ein virtueller Speicher für E-Mails, Photographien, Adressen und Termine, der diese Daten auch mit dem iPad und Apple-Computern synchronisieren kann. An weiteren Apps in Lizenz wie etwa Facebook, Flickr, Myspace, Twitter oder YouTube, digitalen Zeitungsausgaben vom *Stern* bis zum *Spiegel*, von der *Times* bis zur *New York Times*, von der *Bild-Zeitung* bis zur *Welt* verdient Apple mit immerhin 30 % ebenso wie an den commercial Apps, etwa eBay, icq, H & M, Louis Vuitton, Hugo Boss, Dior, Zara, Amazon, New Yorker, Douglas, Chanel oder Tommy Hilfiger, um nur einige zu nennen. Die meisten der genannten Apps sind bei den Smartphones von Nokia und Samsung dagegen kostenlos. Gegenwärtig wächst die Zahl der Apps jeden Tag um ca. 350 neue Anwendungen.

Allein die Anzahl der Apps, die bis zum Erscheinen dieses Buches auf eine Viertel Million steigen dürfte, macht es nicht nur wahrscheinlich, sondern unvermeidlich, daß es eben auch Hunderte von skurrilen und absurden Apps gibt. Das reicht von der Möglichkeit, Seifenblasen über die Displayfläche zu »blasen«, Biertrinken zu simulieren oder das iPhone zum Kerzenausblasen zu nutzen sowie mit dem Mobiltelephon das Geräusch eines Trockenrasierers zu imitieren über gefakte Nutzerprofile von Prominenten wie Barack Obama bis zur Imitation eines »einarmigen Banditen«, also eines Spielautomaten aus Las Vegas, der so tut, als könne er tatsächlich Münzen auswerfen. Es gibt eigene Apps mit Namen wie iPhone Magic oder iPhone Trick. Zudem ermuntern die Apps für elektronische Zeitungsausgaben die Presse zur Werbung in eigener Sache, von der Apple unentgeltlich profitiert. So räumte die *Bild am Sonntag* Mitte 2010 dem iPad eine komplette Doppelseite ein. Abgebildet war ein geöffnetes, auseinandergebautes iPad mit technischen Erläuterungen zu allen Modulen, nicht ohne natürlich auf die Apps des iKiosk vom Axel Springer-Verlag hinzuweisen. So wird die Grenze zwischen Werbung und Berichterstattung fließend: mehr infomercial als infotainment.

228. Advertisement iMac G3, Apple, 1998.
229. Advertisement ibook G3 Clamshell, Apple, 1999.
230. ibeer, Hottrix, 2010.

Road scholar.

 Think different.

catalog of the Zune platform, which comprises eight million songs, for a monthly fee of ten euro. A Zune player, which is similar to the iPod, is available in the United States. Amazon and Deutsche Telekom also offer online music. Even the apps »financial investment« for overnight money and fixed deposits, »real estate finance« with overviews of the appropriate banks and mortgage brokers are also free of charge, while the app »nearby ATMs« costs 1.59 euro, which, however, is always worth the cost given the fact that withdrawing money from banks other than one's own costs quite a bit more. The bank locations are also displayed in 3D.

Apple exclusively and solely delivers the software for music, books, and videos for the iPad as well. This is the best way to keep competitors at bay and to do good business, at least if this software can be distributed for a fee. Among the rapidly growing number of fee-based apps are Apple's online service MobileMe, a virtual storage for e-mails, photographs, addresses and appointments, which can also synchronize this data with the iPad and Apple computers. Apple is generating income – a share of 30 % after all – through licensed apps for Facebook, Flickr, Myspace, Napster, Twitter or YouTube, digital newspaper editions from *Stern* and *Der Spiegel* to *The Times*, *The New York Times*, *Bild-Zeitung* and *Die Welt*, as well as with the commercial apps such as eBay, icy, Louis Vuitton, Hugo Boss, Dior, Zara, Amazon, New Yorker, Douglas, Chanel or Tommy Hilfiger, to name but a few. Most of the aforementioned apps are free, but only for the smartphones from Nokia and Samsung. Currently, the number of apps is climbing by approximately 350 new applications per day.

The number of apps alone, which should climb to a quarter million by the time this book is published, not only makes it possible but a certainty that there will be hundreds of bizarre and absurd apps. The spectrum ranges from the option to »blow« soap bubbles over the display surface to the simulation of drinking beer, to using the iPhone for extinguishing candles or to imitating the sound of a dry shaver with the mobile phone, to fake user profiles of celebrities, like Barack Obama for example, to the imitation of a »one-armed bandit« from Las Vegas that pretends to actually pay out coins. There are also apps with names like iPhone Magic or iPhone Trick. In addition, the apps for electronic newspaper editions encourage the press to advertise on their own, from which Apple profits without doing a thing. *Bild am Sonntag*, for example, dedicated a full double page to the iPad in mid-2010. It showed an opened and disassembled iPad with technical explanations about all of the modules, of course not without pointing out the apps of the iKiosk of Axel Springer Verlag. The boundary between advertising and reporting thus becomes fluid: more infomercial than info-tainment.

But just as electronic media has cannibalized printed products, the print industry in the meantime has set out to »re-Gutenbergize« electronic formats. The press thus reacted to the possibility of downloading magazines and newspapers to smartphones as digital app editions with a counter strategy, which demotes mobile phones to servants of print media. The magazine of *Süddeutsche Zeitung* presented printed pictures in mid-August 2010, which could be animated with the assistance of a smartphone. The necessary technology is called »augmented reality«. In general, it refers to the complementation of images or videos with extra computer-generated information. Applications in sports reporting, construction, maintenance, and medicine, for the military or disaster management, but especially for smartphones are known. The Munich company Metaio, market leader for this technology, animated the magazine edition. With a free app called Junaio users open a channel of the magazine and then hold the mobile phone at a distance of approx. 30 cm on one of the market pages to then view printed pictures that suddenly begin to move. The whole process is a bit reminiscent of the glasses for 3D films. In addition, it sends the impression: The Empire strikes back! While the apps here serve to animate real, haptic printed matter, haptics are suspended in the case of the electronic print version. Mythologies of day-to-day life: Lewis Carroll's fairy tale *Alice in Wonderland* already includes books with animated pictures.

Compared with these printed apps, a first 3D issue of *Bild-Zeitung*, which was rather lavishly advertised, seemed rather harmless. Of course, a pair of the required green and red cardboard glasses was included with each copy. This technology has been around for quite a few decades for viewing 3D movies or, more rarely, television broadcasts. It is simply remarkable that Germany's newspaper with the highest circulation was compelled to use this antiquated technology: obviously it was a reaction to Hollywood's successful 3D mainstream movies such as James Cameron's *Avatar*. And the

Aber gerade so, wie die elektronischen Medien die gedruckten Erzeugnisse kannibalisieren, so schickt sich inzwischen die Print-Branche an, elektronische Formate zu »re-gutenbergisieren«. So reagierte die Presse auf die Möglichkeit, Zeitungen und Zeitschriften als digitale App-Ausgabe auf das Smartphone zu laden, mit einer Gegenstrategie, die das Mobiltelephon umgekehrt zum Diener der Printmedien degradiert. Das Magazin der *Süddeutschen Zeitung* präsentierte Mitte August 2010 gedruckte Bilder, die mit Hilfe eines Smartphones animiert werden können. Die Technik dafür nennt sich »Augmented Reality« (»erweiterte Wirklichkeit«). Im allgemeinen versteht man darunter die Ergänzung von Bildern oder Videos mit computergenerierten Zusatzinformationen. Anwendungen in der Sportberichterstattung, der Konstruktion, Wartung und Medizin, beim Militär oder dem Katastrophenmanagement, vor allem aber auch bei Smartphones sind bekannt. Die Magazinausgabe hat die Münchner Firma Metaio, Marktführer für diese Technologie, animiert. Mit einem kostenlosen App namens Junaio öffnet man einen Kanal des Magazins und hält dann das Handy mit etwa 30 cm Abstand auf eine der jeweils gekennzeichneten Seiten. Dann sieht man gedruckte Bilder, die sich plötzlich bewegen können. Das Ganze erinnert ein bißchen an die Brillen für 3D-Filme. Zudem drängt sich der Eindruck auf: The Empire strikes back! Denn während die Apps hier zur Animierung einer haptisch realen Drucksache dienen, wird die Haptik bei der elektronischen Druckversion suspendiert. Mythologien des Alltags: Schon in Lewis Carrolls Märchen *Alice im Wunderland* gibt es Bücher mit bewegten Bildern.

Gegenüber diesen gedruckten Apps wirkte dann ein paar Wochen später eine aufwendig beworbene, erste 3D-Ausgabe der *Bild-Zeitung* eher harmlos. Natürlich wurde jeder Ausgabe die unvermeidliche Grün-Rot-Pappbrille beigelegt. Diese Technik ist seit mehreren Jahrzehnten für Kinofilme oder, seltener, auch für TV-Ausstrahlungen bekannt. Immerhin ist bemerkenswert, daß nun die auflagenstärkste Tageszeitung Deutschlands diese Technik aufgreift: offensichtlich ein Rückkopplungseffekt zu dem seit kurzen reüssierenden 3D-Mainstream-Spielfilmen Hollywoods wie etwa *Avatar* von James Cameron. Und das Mail Order-Produktunternehmen Ikarus übernahm in seinem Katalog jene kleinen chaotisch punktierten Quadratflächen, die ursprünglich in den 1990er Jahren zur Qualitätssicherung in der Automobil-Zulieferindustrie entwickelt wurden, zur Kennzeichnung mancher Produkte: Über ein iPhone-App kann man sich anschauliche Videos dieser Produkte auf den Touchscreen holen.

Generell sind die Apps oft nochmals unterteilt. Klickt man z. B. Facebook an, dann erscheinen als Unter-Apps die Kategorien Neuigkeiten, Photos, Suchen, Eingang, Ich und Veranstaltungen. Ab Juli 2010 kam iAd, eine eigene Verkaufsplattform für mobile Werbung, hinzu. Dies verspricht zwar in Zukunft ein zweistelliges Milliardengeschäft zu werden, ist aber für die Werbekunden nur dann wirklich interessant, wenn Apple ihnen Zielgenauigkeit, also Werbeeffizienz, ermöglicht. Deshalb öffnet sich ab ca. der Mitte des Jahres 2010 bei jedem Klick auf ein iPod, iPad oder iPhone ein Fenster mit neuen Geschäftsbedingungen, denen die Nutzer zustimmen sollen: Apple speichert, wo wer sich wann mit seinem Gerät aufhält und verlangt außerdem die Zustimmung dafür, diese Informationen weiter verarbeiten und sie an Dritte weitergeben zu dürfen. Neben datenschutzrechtlichen Bedenken, die auch sofort geäußert wurden,[34] bedeutet die Plattform iAd vor allem einen Angriff auf das Kerngeschäft der Online-Suchmaschine Google. Der *Spiegel* schrieb: »Apple gilt als Vorbild in Sachen Datenschutz. Doch tatsächlich will der US-Konzern weltweit wissen, wo seine Millionen Kunden gerade ihre iPhones und iPads nutzen. So läßt sich der Industrie Reklame noch besser verkaufen. Die Attacke zielt vor allem auf Google.«[35] Damit geht der Wettbewerb der beiden Konzerne in die nächste Runde. Schon Googles eigenes Smartphone Nexus mit einem kostenlosen Betriebssystem war ein Angriff auf das iPhone, ebenso ein eigener Musikstore von Google, der iTunes Konkurrenz macht. Generell ist Google mit seinem Betriebssystem Android, welches auch andere Smartphone-Anbieter eingebaut haben, mit mehr als 90 000 Apps der größte Konkurrent für Apple. Es gibt Google web'n'walk und Google talk und andere Google-Funktionen wie Suche, Maps und Mail kann man selbstverständlich ebenso nutzen. Auch bei Google sind Finanzdienstleistungs-Apps wie Currency für mehr als 160 aktuelle Wechselkurse oder Finanzen.net mit Überblicken über europäische Finanzmärkte ebenso wie Kreditrechner für Autos und Immobilien wie Car Loan oder Mortgage Calculator kostenlos. Und mit dem ebenso kostenlosen Tool App Inventor können auch Anfänger einfache Zusatzprogramme entwickeln.

Ende September 2010 meldeten die Agenturen, daß Apple am Drucken per Funk für das iPhone und das iPad arbeitet. Dieses neue App AirPrint soll Teil des Updates für Apples Betriebssystem iOS 4.2 werden. Voraussetzung dafür sind natürlich kompatible Drucker. Im Januar 2011 wurde bekannt, daß Apple – im übrigen auch Google, Amazon und eBay – eigene Zahlsysteme als Industriestandard etablieren wollen. Apple ist mit seinem App Store und dem iBook Store hier Vorreiter und arbeitet bereits an einer eigenen Kioskversion für Verlagsinhalte. Das Zahlen per Mobiltelephon statt Bargeld oder Kreditkarte steht auch auf der Agenda von Vodafone, O2 und der Deutschen Telekom. So testet bereits die Deutsche Bahn mit den genannten drei Anbietern ein Touch & Travel genanntes Fahrkartensystem zum Buchen über Smartphones. Bereits das iPhone 5 ebenso wie Googles neues Nexus S sollen mit der dafür notwendigen Nahfunktechnik NFC (Near Field Communication) ausgerüstet werden.

Und es gehört zur Logik des buchstäblich für alle Inhalte offenen Internets, daß auch die Apps inzwischen skurrile, wenn nicht bizarre Varianten haben. So soll eine spezielle App Akne bekämpfen. Klickt man sie an, dann strahlt das iPhone ein spezielles Licht zur Pickelbekämpfung aus. Ein weiteres App projiziert ein Bildschirm-Laufrad, auf dem unsere Finger joggen sollen, um für die Multi-Gesten-Steuerung fit zu bleiben.

Auch im Islam sind Apps inzwischen willkommen. So bietet etwa Ramadan Daily Dua an jedem Tag des heiligen Monats das passende Gebet an, Ra-

mail order product company Ikarus in its catalog took over the small, chaotically dotted square surfaces that were originally developed for quality assurance in the 1990s automobile industry to mark some products: vivid videos of these products can be loaded onto the touchscreen via an iPhone app.

In general, the apps are often further subdivided. For example, when clicking on Facebook the current sub-apps are, for example, News Feed, Photos, Search, Share, Profile, and Events. In July 2010 iAd, a special sales platform for mobile advertising, was added. It promises to become a multi-billion dollar business in the future, but it is only of interest to advertising clients if Apple offers precise targeting, meaning advertising efficiency. Therefore, since mid-2010, a window with new terms of use opens up each time one clicks on an iPod, iPad or iPhone: Apple stores the location and time and also requests approval to further use this information and to forward it to third parties. Alongside the data protection concerns that were immediately voiced,[34] the platform iAd also represents an attack on the core business of the online search engine Google. *Der Spiegel* wrote: »Apple is considered a model with respect to data protection. But in fact, the US group wants to know worldwide where its millions of customers are currently using their iPhones and iPads. In this way, advertising can be even better sold to the industry. The attack is mainly aimed at Google.«[35] The competition between the two firms is thus entering the next round.

Google's own smartphone Nexus with a free operating system already was an attack on the iPhone, as was Google's music store, which competes with iTunes. Generally, Apple's biggest competitor is Google with its Android operating system, which other smartphone suppliers are using, along with more than 90 000 apps, for their own phones. There is Google web'n'walk and Google talk, and other Google functions such as Search, Maps and Mail can naturally be used as well. Google also offers free financial services apps such as Currency for more than 160 current exchange rates and Finanzen.net with overviews of European financial markets as well as loan calculators for cars and real estate such as Car Loan or Mortgage Calculator. And with the tool App Inventor – also free – even beginners can develop simple add-on programs. At the end of September 2010 the news agencies reported that Apple was working on enabling wireless printing for the iPhone and iPad. This new app named AirPrint is intended to be part of the update for Apple's operating system iOS 4.2. Of course, compatible printers are a prerequisite. In January 2011 the announcement was made that Apple – alongside Google, Amazon, and eBay – plan to establish their own payment systems as an industry standard. With its App store and the iBook store Apple is the pioneer and already working on its own newsstand version for content from publishers. Payment via mobile phone instead of cash or credit card is also on the agendas of Vodafone, O2, and Deutsche Telekom. Deutsche Bahn has been testing a ticket system for booking via smartphones called Touch & Travel in cooperation with the above-mentioned three providers. The iPhone 5 and Google's new Nexus S will be equipped with the required NFC (Near Field Communication) technology.

231. Advertisement iPod Nano chromatic, Apple, 2008.

madan Guide gibt Tipps zum Fasten, iPray gemahnt an die vorgeschriebenen Gebetszeiten und ein weiteres App zeigt via GPS, in welcher Richtung sich Mekka befindet. Auch kann man den gesamten Koran auf sein Mobiltelephon laden und muß ihn so nicht bei jeder Predigt dabeihaben. Im übrigen gibt es sogar Apps für die analoge Welt: z. B. als Magnetset für Kühlschränke oder als Icon-Applikation auf Schränken. Diese unendliche Vielfalt der Apps erinnert an die ähnliche Vielfalt der Klingeltöne für Mobiltelephone, für die es zahlreiche spezialisierte Anbieter gibt: vom Hundebellen und Pferdewiehern über Musik-Jingles bis zum gehauchten »Je t'aime« ist da alles denkbar.

Gegenwärtig sind fünf Betriebssysteme für Smart-phones auf dem Markt: Symbian von Nokia (41,2 % Marktanteil), RIM® von BlackBerry® (18,2 %), Android von Google (17,2 %), iPhone OS von Apple (14,2 %) und ein Betriebssystem von Microsoft (5 %). Der Smartphone-Pionier Apple ist mit seiner Position in dem kleinen, aber lukrativen Oberklassesegment offenbar zufrieden. Marktanalysten glauben nicht, daß Apple in den Massenmarkt einsteigen wird. Apple will nicht Nokia werden. Auch eine Strategie wie bei iPod, also mit verschiedenen Versionen und Preisen den gesamten Markt zu bedienen, wird es beim iPhone nicht geben. Trotzdem kursieren Gerüchte, daß Apple schon Anfang 2011 ein iPhone 5 auf den Markt bringen könnte. Auch mit einem weiteren Netzbetreiber neben dem bisher einzigen in den USA AT&T – vermutet wird Verizon – könnte Apple seinen Marktanteil bei Smartphones erhöhen.

Einmal mehr zeigt sich, daß Apple die Dimension von Märkten eher entdeckt und öffnet als erfindet. Sowohl bei der iPod-Gattung wie bei jenen des iPhones und des iPad machen schlußendlich andere Unternehmen das wirkliche Massengeschäft. Denn trotz aller Faszination an den i-Geräten ist ihr jeweiliger Marktanteil eher bescheiden. So gab bei der Markteinführung des iPhones im Januar 2007 Steve Jobs als Ziel für 2008 an, 10 Millionen iPhones zu verkaufen. Dies entsprach damals gerade einmal einem Prozent des Weltmarktes für Mobiltelephone. Bis April 2010 sind weltweit dann immerhin ca. 50 Millionen iPhones verkauft worden. Das vorliegende Buch ist eher eine designhistorische als wirtschaftswissenschaftliche Untersuchung. Allerdings erscheinen mir einige betriebswirtschaftliche Daten sinnvoll, um die Produktanalysen auch ökonomisch zu fundieren.

In den 1980er und 1990er Jahren hatte Apple öfter mit Problemen zu kämpfen, vor allem, nachdem Steve Jobs in den 1980er Jahren nach internen Konflikten das Unternehmen verlassen hatte. So war z. B. der Newton ökonomisch ein Flop. Erst 1997 kehrte Jobs als Firmenchef zurück (und stellte zügig die Produktion des Newton ein). Ein Jahr später läutete der semitransparente iMac den Wiederaufstieg der Marke Apple ein. Seitdem hat sich der Umsatz von Apple mehr als versechsfacht: von ca. 7000 Millionen Dollar im Jahr 1992 auf knapp 43 000 Millionen Dollar im Jahr 2009. Die Ups and Downs in den Jahresbilanzen zeigen große Schwankungen: 1997 sank das Umsatzwachstum um 28 % bei ca. 1 Million Dollar Verlust, im Jahr 2000 wuchs der Umsatz um 30 %, sank im Jahr darauf erneut um 33 %, stieg im Jahr 2005 um 68 % und im Jahr 2008 um 53 %. Mitte 2010 verzeichnete Apple mit 15,7 Milliarden Dollar den höchsten Quartalsumsatz und mit 3,25 Milliarden Dollar den höchsten Quartalsgewinn seiner Geschichte. Aber solche Mitteilungen sind schon zum Zeitpunkt ihrer Veröffentlichung Geschichte. Denn im letzten Quartal 2010 konnte Apple im Verhältnis zum gleichen Vorjahreszeitraum den operativen Gewinn verdoppeln. Der Umsatz erhöhte sich sogar um 71 % auf 26,7 Milliarden Dollar. An der Börse ist Apple inzwischen mehr wert als Microsoft; mehr noch: Das Unternehmen ist nun der wertvollste Technologiekonzern der Welt. Und wohl nur die Erfindung eines eigenen iPhones könnte die augenblickliche Talfahrt des finnischen Weltmarktführers Nokia beenden, der zwar ständig neue Mobiltelephone entwickelt, aber auf dem lukrativen Markt der Smartphones

And part of the logic of the Internet, which is literally open to all content, is that strange, if not bizarre, variations of apps have appeared. A special app aims to combat acne. If one clicks on it, the iPhone emits a special light for fighting pimples. Another app projects a spinning wheel on which our fingers are supposed to jog to stay in shape for the multi-gesture control. Apps have meanwhile also been welcomed in Muslim countries. For example, Ramadan Daily Dua offers the matching prayer for each day of the holy month, Ramadan Guide offers tips for fasting, iPray provides a reminder of the prescribed prayer times, and another app shows the direction to Mekka via GPS. The entire Koran can be downloaded to one's mobile phone, and users don't have to carry a printed copy. By the way, there are even apps for the analog world: for example as magnet sets for refrigerators or as icon applications on cupboards. This endless diversity is reminiscent of the similar diversity of ring tones for mobile phones, for which numerous specialized suppliers are active in the market: from dog barks and horse whinnies to music jingles and the whispered »Je t'aime«, anything is possible.

Currently, five operating systems for smartphones are on the market: Symbian by Nokia (41.2 % market share), RIM® by BlackBerry® (18.2 %), Android by Google (17.2 %), iPhone OS by Apple (14.2 %), and an operating system by Microsoft (5 %). As the smartphone pioneer, Apple is obviously content with its position in the small but lucrative upper price class segment. Market analysts don't believe that Apple will enter the mass market. Apple does not aim to become Nokia. There also will not be a strategy for the iPhone such as was used with the iPod, i.e., to serve the entire market by offering different versions. Still, rumors are circulating that Apple could deliver an iPhone 5 to the market in early 2011. Apple could also increase its market share for smartphones by opening the iPhone to another network operator in addition to AT&T, which has enjoyed an exclusive deal with Apple in the USA – Verizon is the rumoured suspect.

Once again, we see that rather than inventing them, Apple discovers and opens the dimension of markets. In the end, with regard to the iPod genre and that of the iPhone or iPad, other companies do the real mass business because, despite all of the fascination with the i-devices, their market share is rather modest. When the iPhone was launched in January 2007, Steve Jobs stated that the goal for 2008 was to sell 10 million iPhones. At the time, this equaled just one percent of the global market for mobile phones. But by April 2010, almost 50 million iPhones had been sold worldwide.

This book is more about design history than an economic analysis. However, some economic data seems useful for underscoring the product analyses. In the 1980s and 1990s Apple frequently had to struggle with problems, especially after Steve Jobs had left the company in the 1980s due to internal conflicts. For example, the Newton was a financial flop. Jobs returned as CEO in 1997 (and quickly halted production of the Newton). One year later, the semi-transparent iMac rang in the upsurge of the Apple brand. Since then, Apple's revenues have increased more than sixfold: from approx. $ 7 million in 1992 to almost $43 million in 2009. The ups and downs in the annual financial statements show enormous fluctuations: in 1997 revenue growth had declined by 28 % with a loss of approximately $ 1 million; in 2000 revenues grew by 30 % but dropped again in the following year by 33 %; in 2005 it climbed by 68 % and in 2008 by 53 %. In mid-2010 Apple recorded its highest quarterly revenues with $ 15.7 billion dollars and the highest quarterly profit in its history with $ 3.25 billion. But often such information is already history by the time it is published. In the last quarter 2010/2011, Apple was able to nearly double its operating profit compared with the same period in the previous year, and revenues even increased by 71 % to 26.7

232. Operation system Mac OS X Snow Leopard, Apple, 2009.
233. Browser Safari, Apple, 2007.

weder Apple noch Google, noch Samsung oder LG Paroli bieten kann. Im zweiten Quartal 2008 nahm Apple auf dem weltweiten Computermarkt den sechsten, in den USA den dritten Rang ein. Im gleichen Zeitraum waren 72 % aller verkauften MP3-Player in den USA iPod-Modelle. 85 % aller Songs aus dem Internet in den USA wurden durch den iTunes Store heruntergeladen.

»An diesem Apfel mißt sich die Welt«, schrieb die *Frankfurter Allgemeine Zeitung* und bemerkte: »An der Börse, wo bekanntlich die Zukunft gehandelt wird, ist Apple … meilenweit nach vorne geeilt. Der Kapitalmarkt bewertet den Konzern aus Kalifornien heute mit mehr als 200 Milliarden Dollar. Für das Geld könnte man Daimler und BMW locker zusammen kaufen. Adidas, Die Allianz und sogar die Deutsche Bank gäbe es noch obendrauf.«[36] 1999 stand Apple an 36. Stelle der wertvollsten Marken der Welt, zehn Jahre später auf Platz 20. Solche Erfolge liegen nicht nur an dem Gestaltungsrenommee des Unternehmens, sondern vor allem daran, daß die i-Produkte eben einen Kultcharakter haben, der manchmal die eingangs erwähnte »Objektophilie« tangiert.

Diesem Kultcharakter scheint auch eine so beträchtliche Anzahl gestohlener iPhones geschuldet zu sein, daß Apple sich eine Erfindung gegen diese Diebstähle hat patentieren lassen. Dieser Diebstahlschutz namens Find My iPhone wird im Rahmen der kostenpflichtigen Apps MobileMe angeboten. Dafür nutzt diese App das im iPhone eingebaute GPS-Modul, mit dem sich die Position des gestohlenen oder verlorengegangenen Gerätes feststellen läßt. Die sensible Erkennungssoftware bemerkt sofort, daß das iPhone entwendet wurde, weil ihr Gesicht, Herzschlag oder Stimme fremd vorkommen. Dann sendet das gestohlene Gerät automatisch ein Photo des Diebes sowie den Aufenthaltsort des Gerätes via Google-Map und übermittelt diese Daten ohne Wissen des Diebes an den Computer des Besitzers, eine Apple-Website, an die Polizei oder andere Behörden. Einem vergleichbaren Sicherheitsbedürfnis von Apple dürfte geschuldet sein, daß dieses Unternehmen neuerdings bei allen seinen Geräten sonderangefertigte Schrauben verwendet. So kann weder ein Nutzer noch ein nicht autorisierter Händler Veränderungen an den Geräten vornehmen.

In der historischen Dimension aber wollen wir doch noch einmal mit Peter Sloterdijk an die lange Entwicklung der Tele-Techniken erinnern: » Das Präfix Tele zeigt jedes Mal an, daß bei steigendem Medieneinsatz die Interaktionen zwischen Akteuren im Begriff sind, den Charakter einer lokalen Begegnung zu verlieren. Positiv ausgedrückt bedeutet das: In jeder Fernkommunikation wird Ferne in ihrer ursprünglichen Wirkung als trennende Instanz aufgehoben. Der Sinn von Teletechniken liegt eo ipso in der Aufhebung der Trennwirkung von räumlicher Distanz. Durch technische Mittel werden mystische und magische Surrogate des Ferndenkens und Fernhandelns überflüssig gemacht; sie erlauben nun alle möglichen sinnlichen und mentalen Intensionen in absentia durchzuführen: Zu den heute trivial gewordenen Funktionen von Fernhören und Fernsehen (Verstärkungen der beiden natürlichen Fernsinnesleistungen) werden in Zukunft auch gewisse Annäherungen an die Nähesinnesleistungen folgen: Fernriechen, Fernschmecken, Fernfühlen (für das letztere hat Howard Rheingold in seinem Klassiker *Virtual Reality* von 1991 ein berüchtigtes Kapitel der Teledildonik oder Fernmasturbation gewidmet – für die Liebhaber der erotischen Geschmacklosigkeit ein Muß.«[37]

Die Wirkungsmächtigkeit und Reichweite marketingstrategischer Maßnahmen und Effekte beschränkt sich allerdings bei weitem nicht mehr nur auf Printmedien und TV-Spots. Durch die Internet-Aktivitäten verändern sich die Absender- wie die Empfänger-, die Wissens- wie die Zeitstrukturen des Agierens und Reagierens grundlegend. Der überkommene Charakter von Information steht zur Disposition. Außerhalb der Apple-Community ist eher unbekannt, mit welcher Intensität und welchem Umfang die Nutzer von Apple-Produkten Kommentare, Erfahrungsberichte, technische Tips, Kostenanalysen, Verfügbarkeiten bestimmter Gerätetypen, Lob und Kritik ins Internet stellen. Diese virtuelle Kontaktebene umfaßt zehntausende von »Sites«, die ständig aktualisiert, wieder und wieder kommentiert, rekommentiert und »überschrieben« werden. Ein ständiger Strom von Apple-»News« und – »Gossips« kommt hinzu. Diese digitalen Parallelwelten sind tendenziell unendlich und verwandeln das Unternehmen subkutan in eine Nachrichtenagentur. Der tägliche Strom der »News« läßt an Agenturen wie Reuters, UPI oder DPA denken. So entsteht eine eigene »Firmenwirklichkeit«, die anders ist und anderes meint als nur Geräte und Software. Neben die Warenproduktion tritt ein Überzeugungsdiskurs: die Warenbotschaft löst sich vom Warenkörper. Kommunikationstheoretisch ist dies weit mehr als eine umfangreiche Corporate Identity, eher schon Teil einer umfassenden Corporate Culture. Dazu paßt, daß iPhone-affine Kids sich spontan via Facebook zum iPhilosophieren zusammenfinden.

Zunehmend wird diese »News«-Orientierung für große Unternehmen wichtig. Es geht um permanente Kundenmassage, um die Stimulation von Sympathien und das Buhlen um Akzeptanz. Flughafenbetreiber, Airlines oder Bahnunternehmen bringen »good will«-Magazine unters Volk; Automobilfirmen errichten PKW-Erlebnisparks, in denen die Abholung eines Neuwagens zum ganztägigen Freizeitvergnügen camoufliert wird; Energiekonzerne schalten Kampagnen mit knapp bekleideten Schönheiten, und Versicherungen versprechen uns fürsorgliche Vorsorge für alle Lebenslagen.

Die Informationen über die Produkte und Dienstleistungen werden immer schneller getaktet. Sowieso werden Neumodelle von Mobiltelephonen inzwischen im Abstand von vier bis sechs Monaten im Markt eingeführt, aber vielen Benutzern erscheint auch das noch als ein langer, ein zu langer Zeitraum: Was kommt denn Tag für Tag hinzu, was kann mein »electronic device« heute, was es gestern noch nicht konnte? Eine im klassischen Sinne des Wortes Verwirbelung und Besinnungslosigkeit gegenüber den Inhalten und vor allem ihrer Reflexion. Gleichzeitig gefallen sich Universitäten in Produktdauer-Forschungen und Produktlebensalter-Analysen.

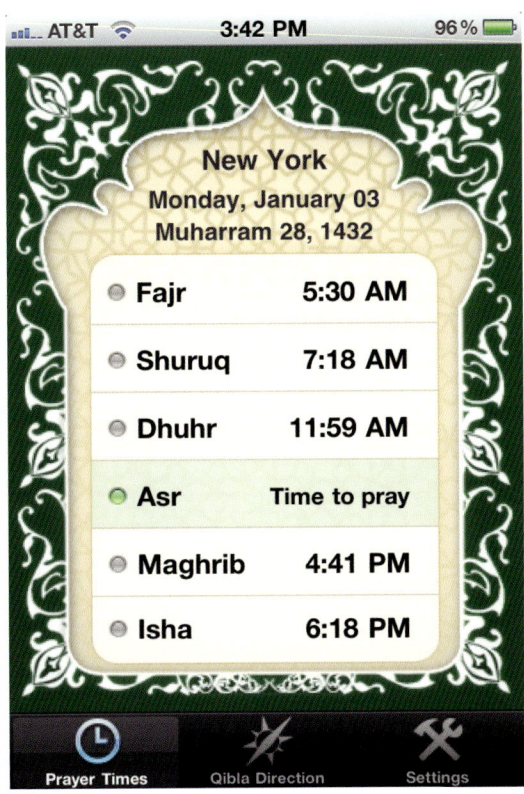

234, 235. App Guidance-Prayer Times, Batoul Apps, 2010.

billion dollars. On the stock exchange, Apple now has a higher market cap than Microsoft; and that's not all: the company has now become the most valuable technology group in the world. Probably only the creation of its own iPhone could stop the current slump of the Finnish global market leader Nokia, which is constantly developing new mobile phones but cannot seem to catch up with Apple, Google, Samsung or LG in the lucrative smartphone market. In the second quarter of 2008 Apple ranked No.6 on the global computer market, and no. 3 in the USA. During the same period 72 % of all MP3 players sold in the USA were iPods. 85 % of all songs downloaded from the Internet in the USA came from the iTunes store.

The *Frankfurter Allgemeine Zeitung* wrote: »The world measures itself against this Apple«, and remarked: »At the stock exchange, where as we know the future is traded, Apple ... is now in the lead by miles. Today, the market values the company from California at more than 200 billion dollars. For this amount of money, Daimler and BMW could be easily purchased. Adidas, Die Allianz, and even Deutsche Bank could be even added as the icing on the cake.«[36] In 1999, Apple ranked No. 36 among the most valuable brands in the world, and ten years later No. 20. Such successes are not only due to the company's design reputation but especially due to the fact that the i-products have a cult character, which sometimes affects the aforementioned »objectophilia.«

One of the results of this cult character is the theft of a considerable number of iPhones. It is such a serious problem that Apple filed a patent for an invention to combat these thefts. This theft protection system called Find My iPhone is offered within the framework of the chargeable app MobileMe. For this purpose, the app uses the GPS module installed in the iPhone to determine the location of the stolen or lost object. The sensitive recognition software immediately notices that the iPhone has been stolen because the carrier's face, heartbeat or voice seem strange to it. The stolen device then automatically sends a photo of the thief and the location of the device via Google Maps and transmits the data, without the thief's knowledge, to the owner's computer, an Apple website, the police, or other authorities. For similar safety reasons Apple recently has developed special designed screws for their devices. So neither a user nor an unauthorized shop is able to change the products.

In the historic dimension, however, we would like to once more remember the long development of telecommunications with Peter Sloterdijk: »The prefix ›tele‹ indicates that the interactions between actors are about to loose the character of a local meeting given the increased use of media. In positive terms, this means: remoteness is annulled in its original effect as a separating instance in each remote or tele-communication. The sense of tele-technics *eo ipso* is to annul the separating effect of spatial distance. Mystical and magical surrogates of tele-thinking and tele-actions become superfluous through technological means; they now allow all possible sensual and mental intentions to be carried out *in absentia*: in the future, the functions of telelistening and television (enhancements of the two natural telesensual performances), which have become commonplace today, certain approaches to the near sensual accomplishments will follow: telesmelling, teletasting, telefeeling (for the latter, Howard Rheingold has dedicated an infamous chapter to teledildonics in his classic Virtual Reality from 1991 – an absolute must for lovers of tasteless erotica).«[37]

The immense power and reach of strategic marketing measures and effects, however, has long ceased being limited to print media and TV spots. The Internet activities have fundamentally changed the sender's and recipient's, knowledge and time structures of acting and reacting. The traditional character of information is up for discussion. The level of intensity and the extent to which the users of Apple products upload comments, field reports, technical tips, cost analyses, availability of specific types of devices, praise and criticism to the Internet is rather unknown outside the Apple community. This virtual realm of contact comprises tens of thousands of sites that are constantly being updated, commented, re-commented and overwritten. A constant flow of Apple news and gossips is added. These digital parallel worlds tend to be infinite, transforming the company into a kind of underground news agency. The daily flow of »news« reminds us of agencies such as Reuters, UPI or DPA creating a special »corporate reality« that is different and means more than just devices and software. A discourse of persuasion now stands beside the production of the product: the product message frees itself from the product body. In terms of communication theory, this is much more than a comprehensive corporate identity; it is rather a part of a comprehensive corporate culture. The fact that kids with an affinity for the iPhone spontaneously gather for iPhilosophizing via Facebook matches this picture.

This »news« orientation is becoming increasingly important for big companies. It is about permanently »massaging« customers, stimulating sympathies and courting acceptance. Airport operators, airlines or railroad companies distribute goodwill magazines; automobile companies establish car event parks where picking up a new vehicle is camouflaged as a full day of pleasure; energy groups launch campaigns with scantily clothed beauties, and insurance companies promise caringly provide for all of life's possibilities.

The information about the products and services is moving at an increasingly faster pace. New mobile phone models are being introduced to the market within four to six months, but many users seem to consider even that too long: what is added day after day, what can my »electronic device« do today that it was not able to do yesterday? In the classic sense, this is a turbulent whirlwind of unconsciousness towards the content and especially its reflection. At the same time, universities please themselves with product life research and product life cycle analyses.

Epilog: Ansichten, Einsichten, Aussichten

Es sollte deutlich geworden sein, daß die i-Geräte den Markt der Unterhaltungselektronik gravierend verändert haben. Sie haben das mobile Internet in Schwung gebracht, digitale Musik, Filme und das Surfen mobil gemacht; wie Villém Flusser sagen würde, »entortet«. Vor allem aber haben sie die Gewohnheiten und die Möglichkeiten der Nutzung solcher Geräte verändert und erweitert. Mehrmals am Tag aktualisierte Zeitungen, Spielfilme »on demand« und Tausende von individuell zusammengestellten Songs immer und überall abrufbar zu haben, verändert nicht nur das Freizeitverhalten, sondern auch das Verhältnis von Freizeit und Arbeit überhaupt, denn beides wächst, nicht zuletzt auch durch solche Geräte, zu- und ineinander.

Es ist gerade einmal zehn Jahre her, daß mit der Versteigerung der UMTS-Lizenzen in Deutschland die Ära des mobilen Internets begann. Eine Handvoll Netzbetreiber zahlten damals 50 Milliarden Euro. Um die hohen Investitionen wenigstens teilweise zu amortisieren, verlangten sie über Jahre sehr hohe Datentarife. Auch war es vom Handling her nicht ganz einfach, sich über ein Mobiltelephon ins Netz einzuloggen. Erst mit dem BlackBerry® (ab 2002) und dem iPhone (ab 2007) sowie inzwischen vielen weiteren Smartphones, vor allem aber auch dem iPad wird das mobile Internet gegenwärtig zum Massenmarkt. Ende 2007 nutzten in Deutschland 8,5 Millionen Teilnehmer regelmäßig UMTS, Ende 2009 schon mehr als 19 Millionen. Im April und Mai 2010 fand eine weitere UMTS-Lizenzversteigerung statt. Besonders begehrt waren jene Frequenzen, die durch die Umstellung auf digitalen Rundfunk nicht mehr genutzt wurden. Im Gegensatz zu dem Hype im Jahr 2000 lag das Auktionsergebnis bei gerade einmal etwas mehr als vier Milliarden Euro. Dies entspricht etwa einem Achtel der damals erzielten Summe.

Die i-Geräte haben, wie gesagt, jeweils ihre Produktgattungen so verändert, daß alle Mitbewerber sich an ihnen orientieren: Insofern haben sie auch neue Wachstumsmärkte geöffnet. Und nur nebenbei: Auch für Kommunikationsdesigner ist es neu, spannend und anders, neben Produktauftritten im stationären Internet mobile Portale zu gestalten. Ein weiteres: Die berührungssensitiven Touch-Displays verändern die Gewohnheiten der Haptik, der Optik und Visualität. Visualität verstanden als die Dialektik von Begreifen und Sehen, von Repräsentation und Perzeption, also als ein Changieren zwischen Zwei- und Dreidimensionalität. Das eine geht im anderen auf, und das eine bildet sich im anderen ab. Neben der Dimensionsfrage gibt es im Kontext der Visualität die Distanzfrage von nah und fern. Neben allen Abstraktionsüberlegungen, die dieses Distanzverhältnis immer und zuallererst betreffen, wird im Falle der i-Geräte das Vokabular dieser räumlichen Distinktionen gehörig verwirbelt, nicht nur im räumlich realen, sondern auch im simulierten räumlichen Sinn.

Pragmatisch gewendet, bleibt festzuhalten: Längst gibt es z. B. Versuche mit aufrollbaren Bildschirmen mit einer Dircke von wenigen Millimetern. Insofern können wir in ein paar Jahren nicht nur mit T-Shirts rechnen, auf denen Nachrichten und bewegliche Bilder zu sehen sein werden, sondern wohl auch mit einem aufrollbaren, internetfähigen iPad. Schon vor etwa 15 Jahren reüssierte der Begriff wearable electronics. In Ohrringen oder Broschen werden interaktive elektronische Funktionen integriert, faßt Notruf-Aggregate für Senioren. Ebenso gibt es längst elektronische Bilderrahmen, auf deren Bildflächen beliebige Photographien zu programmieren sind. Damit wird ebenso eine professionelle Technik massenkompatibel wie etwa eine Software für Laien zum Selbstdrucken und -layouten von Photoalben.

Ob Apple das von Sloterdijk prognostizierte »Fernfühlen« jemals in einem Gerät realisieren kann oder will, bleibt abzuwarten. In jedem Fall aber geht die Miniaturisierung und Nanotechnik weiter. Auch über elektronische Implantate wird bereits nachgedacht. Schon 1988 stellte die Düsseldorfer Designergruppe »Kunstflug« als Konzeptstudie ihren »Fingerelektronischen Handrechner« vor, der Zahlen in die Fingerkuppen und Handflächen implantiert, um mit entsprechenden Bewegungen der Finger zueinander rechnen zu können.[38]

Wenn wir sich heute abzeichnende Konvergenzentwicklungen in der Unterhaltungselektronik in Rechnung stellen, scheinen folgende Produktentwicklungen für Apple durchaus wahrscheinlich:

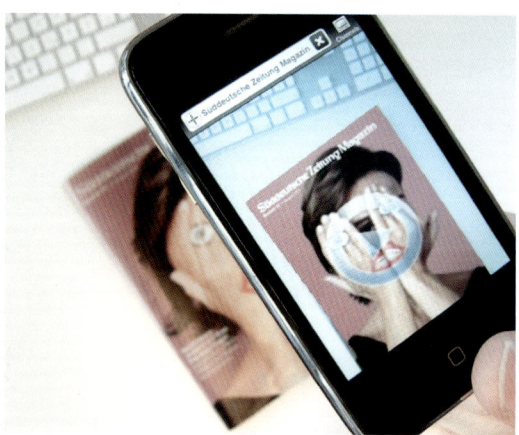

236, 237. Augmented Reality Browser Junaio, Metaio, 2010.
238–240. frog design: Augmented Reality – Envisioning Your Future in 2020, 2010.

Epilogue: Views, insights, outlooks

By now it should be clear that the i-devices have strikingly changed the entertainment electronics market. They have boosted the mobile Internet, and made digital music, films, and Internet surfing mobile; as Villém Flusser would say: »delocalized.« But, above all, they have changed and expanded the habits and possibilities of the use of such devices. Having newspapers that are updated several times a day, films on demand, and thousands of individually compiled songs available anyplace, anytime, not only changes leisure time behavior but also the relationship between leisure and work because the two – not least also due to such devices – are growing closer and merging into one another.

It has only been ten years since the auctioning of the UMTS licenses in Germany rang in the era of the mobile Internet in Germany. At the time, a handful of network operators paid a total of 50 billion euro for the licenses. In order to depreciate the high investments at least in part they charged very high data tariffs for years and, with regard to usability, it was not easy to log onto the web via a mobile phone. The mobile Internet is currently becoming a mass market only because of the BlackBerry® (starting in 2002) and the iPhone (starting in 2007) as well as many other smartphones, but mainly also the iPad. By the end of 2007 8.5 million participants were regularly using UMTS in Germany, and by the end of 2009 more than 19 million. In April and May 2010 another UMTS license auction took place. Those frequencies that were no longer used due to the changeover to digital broadcasting were especially coveted. Contrary to the hype in the year 2000, the auction's result was just a bit over four billion euro. This equals approximately an eighth of the formerly achieved sum. As mentioned, the i-devices have altered their product genres to the extent that all competitors use them for orientation: in this regard, they have also opened new growth markets. And, as a side note: for communication designers it is new, exciting and different to design mobile portals in addition to product presentations on the stationary Internet.

One more thing: the touch-sensitive displays change the habits of haptics, optics, and visuality. Visuality understood as the dialectics of grasping and seeing, of representation and perception, as a changing between two- and three-dimensionality. One is dissolved in the other and one is reflected in the other. In addition to the question of dimension, the distance question of near and remote exists in the context of visuality. Alongside all considerations of abstraction that always and first and foremost concern this distance relation, the vocabulary of this spatial distinction is thoroughly swirled in the case of the i-devices, not only in the spatially real but also in the simulated spatial sense.

Pragmatically speaking, we need to state: there have long been, for example, tests of prototype roll-up screens with a thickness of just a few millimeters. In this regard we can not only count on T-shirts that will display news and moving images but also a roll-up Internet-compatible iPad a few years from now. About fifteen years ago the term

– Da eines der dominanten Merkmale der i-Geräte ihre Mobilität ist, sind erweiterte Kooperationen mit technischen Mobiltätsträgern wie Automobilen und Zügen, Flugzeugen und Schiffen evident. Personalisierte Videodisplays in Auto-, Bahn- und Flugzeugsitzen gehören in den gehobenen Klassen längst zum Standard. Schon heute gibt es zahlreiche Autoradios mit mobilem Internetzugang via iPod oder iPhone. Nicht nur fast alle amerikanischen Automobilunternehmen, sondern auch Peugeot, BMW und Audi bieten mobiles Surfen. Neben dem iPod und dem iPhone ist nun auch das Touchpad im Auto angekommen. Der neue Audi A8 hat über dem Kardantunnel ein nahezu quadratisches Touchpad eingebaut, welches erstmals mobilen Internetzugang während des Fahrens ermöglicht. Dies funktioniert durch Handschrifterkennung. Man legt die Hand auf den großen Wählhebel des Automatikgetriebes und hat damit geradezu automatisch Fingerkontakt zu dem Touchpad, welches Audi MMI Touch nennt. Der Zeigefinger malt nacheinander Buchstaben auf die Fläche, ohne auch nur einmal den Blick von der Straße nehmen zu müssen. Das erkannte Wort oder eine bei Google nachgeschlagene Adresse wird vom System noch einmal leise vorgesprochen. Erste Erfahrungsberichte sind positiv. Noch umfassender hat das deutsche Unternehmen Brabus die Verbindung von Automobil und i-Geräten intensiviert. Auf der Basis eines getunten Mercedes der S-Klasse hat die Firma eine luxuriöse Chauffeurslimousine entwickelt, bei der alle Multimediafunktionen wie Navigation, Musikanlage und Telephon sowie die Innenbeleuchtung und die Klimaanlage über ein iPhone von den Passagieren im Fond gesteuert werden können. An den Rückenlehnen der Vordersitze befinden sich zudem ausklappbare iPads, die sich auch außerhalb des Wagens nutzen lassen. Als Rechenzentrale dieser Anlage für 48 000 Euro dient ein Apple-Notebook in einem speziellen Fach im Kofferraum. Das System wurde erstmals im Spätsommer 2010 auf dem Autosalon in Moskau vorgestellt. Bei solchen parvenuhaften Aufrüstungen bleibt der tatsächliche automobile Nutzen allerdings zweifelhaft. Nicht ausgeschlossen, daß früher oder später ein Konstrukteur vorschlägt, Steuerung und Bremsverhalten über die i-Geräte anzuwählen.

– Zum Kontext der Mobilität gehört auch der Sport als eine dominante Mentalitätssignatur der Gegenwart. Ähnlich wie bereits mit Nike könnte Apple in Zukunft verstärkt in diesem Sektor tätig werden: Smart clothing, also elektronisch optimierte Kleidung, ist ein Markt der Zukunft; – und wie wir am Beispiel von Urban Tool gesehen haben, bereits teilweise Realität.

– Auch die Gesundheit ist so eine Mentalitätssignatur. Warum also sollte es nicht etwa Blutdruck-, Blutzucker- oder Kalorien-Meßgeräte mit iTunes-Programmen geben oder eingebaute Docking Stations in Fitness-Standfahrrädern?

– Kontrollgeräte für Körperfunktionen, wie z. B. Herzschrittmacher, werden seit längerem bereits subkutan implantiert. Vorstellbar ist, daß in Zukunft solche Implantate mit Entertainment-Funktionen ergänzt werden: Dann ginge uns möglicherweise der iPod buchstäblich »unter die Haut«: ein iPod skin?

– Die i-Geräte werden zunehmend auch andere Produktgattungen des privaten Haushalts wie Leuchten, Küchen- und Personenwaagen, Eierkocher und Radiowecker erobern. Einige solcher sekundären i-Peripherieprodukte gibt es bereits. Offenbar ist der i-Kosmos in Hinsicht auf »connectivity« tendenziell gattungsübergreifend.

– Auch Gebrauchsanweisungen als akustische Menüs wären schon deswegen attraktiv, weil man

241. iDrive, BMW, 2008.
242, 243. MMI control panel, Audi. 2009.
244. KIV-700, Kenwood, 2010.
245. GPS navigation system nüvi 3790T, Garmin, 2010.
246–248. iBusiness, Brabus 2010.

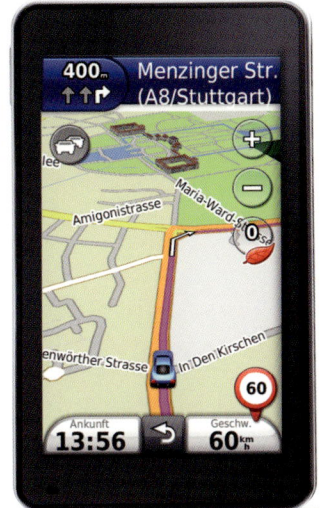

wearable electronics was successfully created. Interactive electronic functions are integrated in earrings or brooches, for example emergency call systems for seniors. Also, electronic picture frames with displays that allow users to program random photographs have long been available. This makes a professional technology mass-market compatible, as much as laymen's software for the print-out and layout of photo albums does.

It remains to be seen whether Apple will ever be able and willing to realize the »telefeeling« that Sloterdijk predicted. At any rate, however, miniaturization and nano-technology will progress. Considerations about electronic implants are already underway, as well. As early as in 1988 the Düsseldorf designer's group »Kunstflug« already presented its so-called »finger-electronic calculator« as a concept study, which implants numbers into the fingertips and palms in order to be able to calculate through the appropriate finger movements.[38]

If we take convergence developments in entertainment electronics that are emerging today into account, the following product developments in a way seem probable for Apple:

– Since mobility is one of the dominant characteristics of the i-devices, extended cooperations with technological mobility carriers such as automobiles and trains, airplanes and ships are evident. Personalized video displays in car, train, and airplane seats have long become a standard in the upper price segment. Today, numerous car radios already include mobile Internet access via iPod or iPhone. Not only do almost all US American automobile brands offer mobile surfing, but also Peugeot, BMW, and Audi. Now, alongside the iPod and the iPhone, the touchpad has arrived in cars. The new Audi A8 has an almost square touchpad installed above the transmission tunnel which for the first time enables mobile Internet access while driving. It functions through handwriting recognition. You place your hand on the big gear shifter and thus almost automatically have finger contact with the touchpad, which Audi calls MMI Touch. The driver scrawls letters on the surface with the index finger without having to take his or her eyes off the road even for a moment. The recognized word or address looked up in Google is then softly spoken by the system. Early field reports have been positive. The German company Brabus has intensified the combination of automobile and i-devices even more. Based on a tuned Mercedes S-class, the company has developed a luxurious limousine in which all multi-media functions such as navigation, music system and phone as well as interior lighting and air conditioning can be controlled by the passengers in the back via an iPhone. In addition, fold-out iPads are mounted on the backrests of the front seats that can also be removed and used outside the car. An Apple notebook in a special compartment in the trunk, which costs 48 000 euro, serves as the computer center of this system. The system was first presented at the automobile salon in Moscow in the summer of 2010. However, the actual automobile use remains dubious in the case of such parvenu-like upgrades. It is entirely possible that sooner or later an engineer will suggest controlling the steering and braking via the i-devices.

– Sports also belong to the context of mobility as a dominant mentality signature of the present. Similar to Nike, Apple could become increasingly active in this sector in the future: smart clothing, electronically optimized clothing, is a market of the future – and as we have seen on the example of Urban Tool, this is already reality.

– Health, too, is a mentality signature. So why shouldn't there be a blood pressure, blood sugar or caloric measuring device with iTunes programs or docking stations integrated in stationary fitness bikes?

– Devices such as pacemakers to control physical functions have been subcutaneously im-

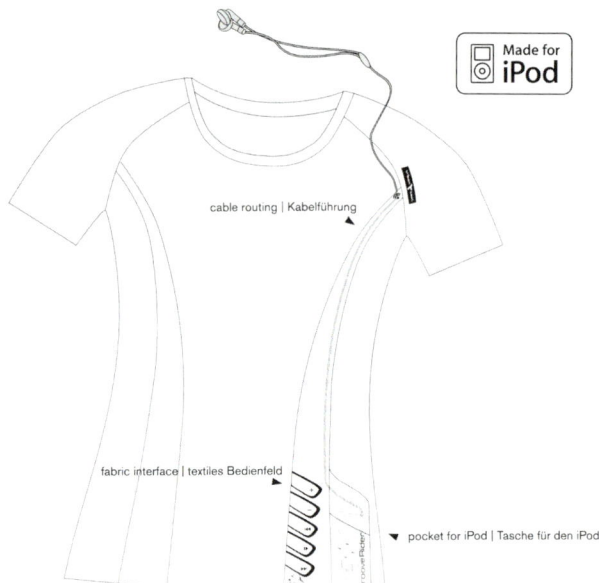

sie überall und vor allem im Freien abhören kann. Vom Motorrad bis zum Rasenmäher, vom Außenkamin bis zum Luxusgrill würde man Schritt für Schritt während der Handhabung in eben diese eingeführt.

– Spätestens ab 2025, so die Prognosen, wird es möglich sein, die Wahrnehmungen aller Sinne – sehen, hören, schmecken, riechen, tasten – künstlich zu verstärken bzw. zu konfigurieren. Implantate in Ohren und Augen werden durch die Nanotechnologie zu einem Massenmarkt werden. Augen-Apps, z. B. auf die Netzhaut projizierte Entfernungen, Stadtpläne oder Wetterdaten sind ebenso denkbar wie Ohren-Apps, also Musik oder akustische Orientierungen im Innenohr. Längst sind ja solche Techno-Utopien in Science-fiction-Filmen imaginiert und antizipiert. Also: iEye, iEar, iNose, iSmell, iFeel?

– Wahrscheinlich wird Apple und die i-Family früher oder später auch im Bereich der Virtual Reality aktiv werden. Die Internetplattform Second Life, eine vom Benutzer bestimmte Parallelwelt, auch 2.0 genannt, weist bereits in diese Richtung. Auch Spielkonsolen wie Nintendos Wii oder neuerdings das Kinect-System von Microsoft oder Sonys Move »übersetzen« eine Realität (die der realen Bewegung) in eine andere (die des Bildschirms). Kinect ist eine schwarze Box, die am Fernsehgerät angebracht wird. Kameras und Infrarotsensoren erfassen die Bewegungen von ein oder zwei Spielern und übertragen diese an die Konsole. Bei Sonys Move dagegen muß der Spieler zwei Eingabegeräte in der Hand halten. Und 3D-Animationen, inzwischen im Mainstream der Hollywoodspielfilme angekommen, erobern gegenwärtig auch die Spielkonsolen. Schon drängen die Konsolenspiele auch auf das iPhone und das iPad. Denkbar ist insofern auch, daß Apple irgendwann selbst wie Nintendo oder Sony eigene mobile Spielkonsolen anbietet. Die Spielkonsolen der Zukunft jedenfalls werden wohl alle 3D-animiert sein.

– Auch im Bereich der »Augmented Reality« könnte sich Apple engagieren. Gewissermaßen mit einem Salto rückwärts iPad-Logistik in Printmedien zu reimplantieren, mag sich als ökonomisch lohnend herausstellen. Auch wird man virtuelle Geräte im realen Raum darstellen können, die durch Blicke oder Fingerbewegungen bedient werden. Möglicherweise werden auch die Displays der Smart-phones und Tablet-Computer durch Informationseinblendungen direkt in die Umwelt ergänzt oder ersetzt werden.

– Seit gut zehn Jahren werden private Bäder zu Wellness-Tempeln aufgewertet. Ich habe diese generelle Entwicklung einmal als »Vom Abtritt zum Auftritt« bezeichnet. Selbst das Kultunternehmen für trendigen Küchen-Lifestyle, Alessi, vermarktet seit 2006 eine eigene Kollektion von Sanitärkeramik. Bei öffentlichen Toiletten in Japan und den USA wird ambient music eingespielt, um Körpergeräusche zu übertönen. Auch da sind stationäre i-Funktionen denkbar.

– Es gibt bereits Hotels, die ihre Zimmer dominant mit deren iPod-Kapazitäten vermarkten. Vor ein paar Jahren waren das noch die Internet-Steckdosen für mitgebrachte Laptops. Heute kann man über entsprechende »Docking Stations« in diesen Hotelzimmern mobil alle Internetinhalte abrufen. Daß Hotels überhaupt mit einem solchen Angebot werben, verweist auf den Status der i-Geräte. Offenbar sind sie dabei, als touristisches Werbepotential für Beherbergungsangebote ebenso selbstverständlich zu werden wie für TV-Geräte, Föne und Telephone. Weitergedacht: Statt läppischen Welcome-Botschaften auf dem TV-Bildschirm könnte man via iPod wichtige Informationen über Kulturangebote, Gastronomie und Sehenswürdigkeiten der entsprechenden Region als Apps abrufbar halten. Die Frankfurter Lokalzeitung *Neue Presse* stellte bereits regionale »Apfelwein«-Apps vor.

– Dies führt zu den sog. Intelligent Buildings, die bereits bauseits elektronische Steuerungssysteme

planted in patient's bodies for quite some time. It is conceivable that such implants will be complemented with entertainment functions in the future: then the iPod might literally get »under the skin«: an iPod skin?

– The i-devices will increasingly conquer other product genres in households, such as lamps, kitchen or personal scales, egg boilers and radio alarms. Some of these secondary i-peripheral devices already exist. Obviously, the i-cosmos is cross-genre with regard to connectivity.

– Instruction manuals as acoustical menus would be attractive just because of the fact that they can be listened to anyplace and, above all, outdoors. From motorcycles to lawn mowers, outdoor chimneys to luxury barbecues – one would be introduced to them step by step during operation.

– It is projected that by 2025 at the latest it will be possible to artificially enhance or configure the perceptions of all senses – seeing, listening, tasting, smelling, touching. Implants in ears and eyes will reach the mass market due to nano technology. Eye apps, e. g., distances, city maps or weather data projected onto the retina are conceivable, as are ear apps, i.e., music or acoustic orientations in the inner ear. Such techno-utopias have long been anticipated in science fiction books and movies. Therefore: iEye, iEar, iNose, iSmell, iFeel?

– Apple and the i-Family will probably become active in the virtual reality sector sooner or later. The online platform Second Life, a parallel word defined by the user and also called 2.0, already points in this direction. Games consoles such as Nintendo's Wii or recently Microsoft's Kinect system or Sony's Move »translate« a reality (that of real movement) into a different reality (that of the monitor or display). Kinect is a black box that is connected to the television set. Cameras and infrared sensors capture the movements of one or two players and transmit them to the console. In the case of Sony's Move, however, the player has to hold two input devices in his or her hands. 3D animations, which have meanwhile become part of the mainstream in Hollywood, are currently also taking over games consoles. Already, these console games are moving onto the iPhone and iPad. In this respect, it is also conceivable that Apple will at some point – like Nintendo or Sony – offer its own mobile games console. Anyway, the games consoles of the future will most likely all be 3D animated.

– Apple could also engage in the field of »augmented reality«. Reimplanting iPad logistics in a sense with a backwards somersault may prove to be economically worthwhile. Virtual devices will also be represented in real space, and they will be operated through visual contact or finger movements. Possibly the displays of the smartphones and touch pads will be complemented or replaced through information fade-ins directly into the environment.

– For more than ten years, private bathrooms have been upgraded into wellness temples. I once described this general development as »from exit to entry«, using a picture from the world of theater. Even Alessi, the trendy kitchen lifestyle company, has been marketing its own collection of sanitary ceramics since 2006. Ambient music is played in public restrooms in Japan and the USA in order to mute body sounds. Here, too, stationary i-functions are a possibility.

– There are already hotels that strongly market their rooms as being iPod ready. A few years ago, it was Internet outlets for laptops. Today, Internet content can be accessed via the appropriate docking stations in hotel rooms. The fact that hotels advertise such offers at all indicates the status of the i-devices. Obviously, they are about to become as self-evident as TV sets, hairdryers and telephones as advertising potential for lodging of-

249, 250. grooveRider, Urban Tool, 2006.
251. Nike+iPod Sport Kit, Apple 2006.
252. App Nike+, Nike, 2010.
253. Wristband Nike+, Nike, 2010.

für Licht, Wärme, Lüftung, Verdunkelung, Sicherheit, Beschallung und TV-Bildwände eingebaut haben. Auch mobile funkgesteuerte Fernbedienungen für Gebäudefunktionen sind längst auf dem Markt, in Deutschland z. B. von Siedle oder Merten. Video-Kennung über Miniaturkameras zur Eingangskontrolle gehört heute schon zur Normalausstattung von Eigentumswohnungen im gehobenen Preissegment. Das Licht »an- und auszusprechen«, hat in manchen Apartments längst die Wandschalter ersetzt. Was läge also näher, als solche smart buildings über Smartphones anzusteuern? iPhone-Apps für Gebäudesteuerungen werden nicht lange auf sich warten lassen, wenn sie nicht längst schon vorhanden sind. Und warum sollte es nicht irgendwann flächenbündig eingebaute iPads von zwei auf drei Meter geben?

– Auch in den Bereichen Pädagogik und Didaktik werden die i-Geräte verstärkt genutzt werden. Schon heute etwa setzt das Grassi-Museum für Angewandte Kunst Leipzig versteckt eingebaute iPods als elektronische Museumsführer ein, die über Dauerschleifen laufen und jeden Morgen von den Aufsichten per Funk aktiviert werden. Pädagogische Nutzungen sind auch im Sprachunterricht, nicht zuletzt bei Nutzern mit Migrationshintergrund, denkbar und sinnvoll.

– Sogar künstlerische Nutzungen des iPads sind bereits Realität. So hat der New Yorker Künstler David Jon Kassan mit der Brushes-App innerhalb von drei Stunden nur mit seinen Fingern das naturalistische Porträt eines alten Mannes gemalt, welches tatsächlich wie ein Leinwandbild aussieht. Auch der englische Maler David Hockney benutzt das iPad als digitale Leinwand, wobei seine elektronischen Bilder eher an Graffiti erinnern. Insgesamt scheinen solche künstlerischen Nutzungen zuzunehmen. Die Verwendung von i-Geräten in interaktiven Kunstinstallationen scheint nur eine Frage der Zeit zu sein, zumal diese Interaktivität ja durchaus technische Vorläufer hat – von den TV-Geräten Nam June Paiks über die Erkennung von »Handschriftlichem« auf elektronischen Displays, wie etwa im Mozarthaus in Salzburg bis zu elektronisch reproduzierten Ölporträts, bei denen man durch Berührungen der Displayfläche die Vorzeichnung nach und nach sichtbar machen kann. Verallgemeinert: Die digitalen Displays und die sie speisenden Software-Programme werden technisch immer komplizierter und leistungsfähiger, benutzungsstrategisch dagegen immer einfacher und selbstevidenter. Insofern sind die Benutzeroberflächen der Maschinen nicht nur, wie Sloterdijk bemerkt, ihr Make-up (denn damit wird der Natur ja künstlich nachgeholfen), sondern sie ermöglichen in einer bemerkenswerten Volte gerade durch die Verwendung avanciertester Software vorindustrielle »handwerkliche« Techniken.[39] Mit »Fingermalen« assoziiert man unwillkürlich frühkindliche Sozialisation. Auf dem iPad haben solche Verfahren deshalb etwas Regredierendes, weniger in technischer als in sozialpsychologischer Hinsicht und im Hinblick auf Medienkompetenz. Insofern ist diese Brushes App die Höhlenmalerei der digitalen Zivilisation: als wenn die Fels/Kohle-Zeichnungen aus der Höhle von Lascaux plötzlich ins Silicon Valley gebeamt wären.

Jedenfalls werden in der Unterhaltungselektronik im nächsten Jahrzehnt die drei »M« »Mobilität, Miniaturisierung und Mainstream« ebenso bestimmend bleiben wie die drei »K« »Konvergenz, Kompatibilität und Kommunikation«.

Im Zusammenhang der erwähnten Veränderungen des Gebrauchs ist eine weitere Überlegung von Peter Sloterdijk erhellend: »(Der) Begriff der »Benutzeroberfläche« (gehört) ... zu den klügsten Wörtern der neueren Umgangssprache. Er deutet auf zwei entscheidende Sachverhalte gleichzeitig hin: zum einen, daß die menschliche Subjektivität in einer technischen Umwelt vor allem auf die Benutzerrolle geprägt ist, ... zum anderen, daß eben dieser Benutzer nur insofern noch Herr und Vorteilsnehmer aus dem Umgang mit technischem Gerät sein kann, als er sich damit begnügt, von der Technik als solcher nichts zu verstehen und sich auf die Bedienung von Tasten und Einstellknöpfen zu beschränken. Die Benutzeroberfläche ist also die Kontaktebene zwischen einem undurchsichtigen Automaten und einer menschlichen Intelligenz, die von vornherein davon dispensiert ist, etwas vom ›Innenleben‹ des Gerätes zu verstehen. ... Die Technokultur hat Erfolg in dem Maß, wie es

fers for tourists. Taking this thought a step further: instead of banal welcome messages on the TV screen, substantial information about cultural offers, gastronomy and sights of the region could be made available via iPod apps. The local Frankfurt newspaper »Neue Presse« has already presented regional »apple wine« apps.

– This takes us to the so-called intelligent buildings, which already have electronic control systems installed for light, heat, air conditioning, shading, security, sound, and TV screens. Mobile remote controls for building functions have long been available on the market, in Germany, for example, by Siedle and Merten. Video recognition via miniaturized cameras for entry control is already standard equipment of apartments in the upper price segment. In some apartments, »speaking« to turn lights on and off has long replaced wall switches. Hence, what would be more at hand than controlling such smart buildings via smartphones? iPhone apps for building controls will be available in the not too distant future, if they are not available already. And why shouldn't there be two by three meter iPads installed flush with any surface?

– The i-devices will also be increasingly used in pedagogics and didactics. Today, for example, the Grassi-Museum für Angewandte Kunst Leipzig uses invisibly installed iPods as electronic museum guides that run via infinite loops and are activated every morning by the museum's supervisors. Pedagogical uses are also conceivable and reasonable in language classes, not least among immigrants.

– Even artistic uses of the iPad have meanwhile become a reality. The New York artist David Jon Kassan painted a naturalistic portrait of an old man with his fingers with the Brushes app in just three hours; it indeed looks like a painting on canvas. English painter David Hockney also uses the iPad as a digital canvas, although his electronic pictures are reminiscent of graffiti. Altogether, such artistic uses seem to be on the rise. The use of i-devices in interactive art installations only appears to be a question of time given the fact that this interactivity absolutely has technical precursors – from the TV sets of Nam June Paik to the recognition of »handscript« on electronic displays, for example in the Mozart House in Salzburg, to electronically reproduced oil portraits in which the sketch can be made successively visible by touching the display. In general terms, the digital displays and software programs feeding them become increasingly complicated technologically and increasingly powerful, but they also become increasingly simple to use. In this respect, the user interfaces of the machines are not only – as Sloterdijk remarks – their make-up (since they artificially give a helping hand to nature) but, in a remarkable volte, they enable pre-industrial »handicraft« techniques especially due to the use of highly advanced software.[39] »Finger painting« is automatically associated with early childhood socialization. On the iPad such methods hence have a regressive aspect, less with regard to technology than socio-psychology and media competence. In this respect, this Brushes app is the cave paint-ing of digital civilization: as though the charcoal drawings from the Lascaux cave suddenly had beamed into Silicon Valley.

At any rate, the three »Ms« in entertainment electronics – »mobility, miniaturization, and mainstream« – will continue to be as defining in the next decade as the three »Cs« – »convergence, compatibility, and communication.«

In the context of the aforementioned changes of use, one more of Peter Sloterdijk's ideas is enlightening: »(The) term ›user interface‹ (is among) … the most intelligent words of the recent colloquial language. It simultaneously points out two decisive facts: first, that human subjectivity in a technological environment is mainly formed into the role of user, … and second, that this very user can only be master and benefactor from the handling of technological devices insofar as he is complacent about not understanding any of the technology as such and is restricted to the operation of keys and adjustment buttons. The user interface is hence the contact level between an opaque machine and a human intelligence that is from the beginning excused from understanding anything about the device's ›inner life‹. …

254, 255. House management system, Jung, 2010.
256. Digital picture frame TFT-1020, Lenco, 2010.
257. Digital picture frame, Hama, 2010.
258. Digital picture frame, Hama, 2010.

ihr gelingt, in den Benutzern die Illusion von Operationsfähigkeit wach zu halten, ohne daß bei ihnen das Unbehagen an der Unmöglichkeit, das Apparatinnere zu verstehen, über eine gewisse Schwelle ansteigt. Dies gilt für alle stationären und mobilen Kommunikationstechnologien und ebenso für das gesamte Umfeld an digitalisierten Maschinen. ... Unweigerlich kommt das Problem der Oberflächen-gestaltung immer dann ins Spiel, wenn (das Gerät) dem Benutzer eine Kontaktseite zuwenden muß, um sich ihm nützlich zu machen, ohne je auf seine undurchdringliche innere Struktur hinzuweisen. Die Benutzeroberflächen sind gleichsam das Make-up der Maschinen; sie simulieren eine Art von Verwandtschaft und Sympathie zwischen Mensch und (Unterhaltungselektronik) ... und flüstern dem Benutzer Initiative ein.«[40]

Die Differenz zwischen Benutzen und Durchschauen sollte uns allerdings nicht erschrecken, betrifft sie doch nicht nur elektronische Geräte, sondern jedes von Menschen geschaffene System von komplexen Produkten und Regelkreisläufen. Wie das Wasser aus dem Hahn und der Strom aus der Steckdose kommt, können in der Regel auch talentiertere Erwachsene kaum erklären. Bei einem Automobil einen Reifen zu wechseln, mag ja noch angehen, aber einen Motorschaden reparieren? Instrumentelle und praktische Intelligenz sind nicht deckungsgleich und waren es nie. Wie sagte Otl Aicher: »Das Greifen ist das Begriffene.« Wir fügen hinzu: »Die Oberfläche der Benutzeroberfläche ist ihre Tiefe.« Manche Kommentatoren und Auguren der Produktkultur halten das iPad für den Zenit des Apple-Kosmos, weil der Markt mit diesen i-Geräten doch irgendwann gesättigt sei (sein müsse). Dem steht entgegen, daß diese Firma inzwischen den Unterhaltungsmarkt wie sonst keine andere global im Griff hat und zudem alle Angebote der Musik-, der Film- und der Informationsbranchen auf den i-Geräten zu installieren sind: von Microsoft und Google über Sony und Nintendo bis zu Warner Brothers und Disney. Insofern steht das Unternehmen für die weitreichendste Vernetzung von Kommunikationsformen, die es jemals gegeben hat. Eine Änderung dieser Vernetzungs- und Konvergenzstrategien ist auf lange Zeit nicht zu erwarten, weil sie ökonomisch, psychosozial, emotional und gewohnheitsstrategisch ohne Alternative sind.[41] Vor allem aber sind die Apple-Produkte libidinös besetzt und zudem gestalterisch perfekter als ihre Konkurrenten. Sie sind in multipler Weise intelligenter: Die i-Welten sind und bleiben wohl sehr lange noch die eigentlichen iQ-Welten der »consumer electronics«.

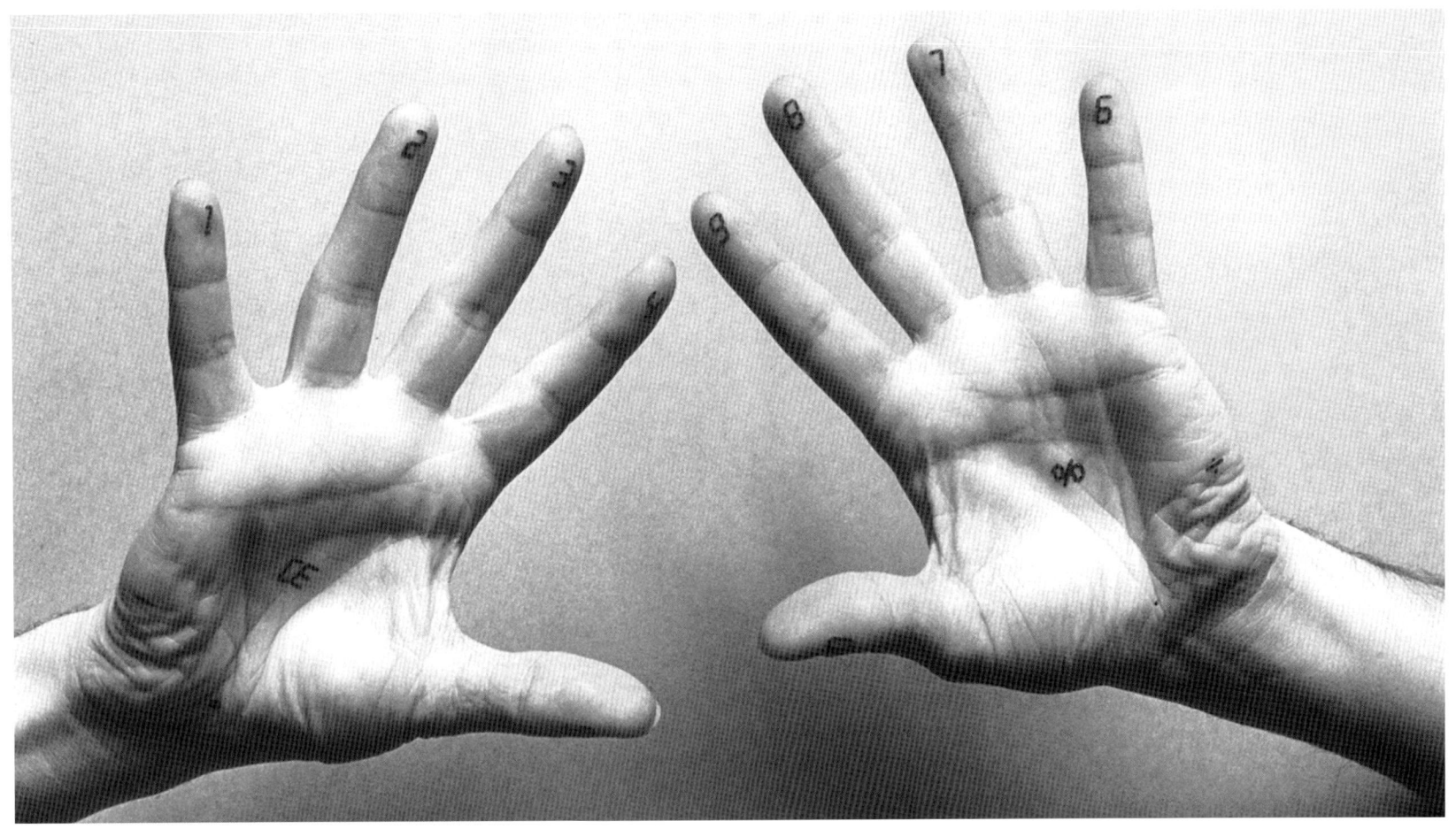

Techno-culture is successful to the degree that it succeeds in keeping the illusion of operational ability alive in the users without their unease about the impossibility of understanding the inner life of the device rising above a certain threshold. This applies to all stationary and mobile communication technologies and also to the entire environment of digitalized machines. ... Unavoidably, the problem of interface design always comes into play at the point when (the device) has to turn a contact side to the user in order to make itself useful without pointing to its impermeable inner structure. The user interfaces are metaphorically the make-up of the machines; they simulate a kind of relationship and sympathy between man and (entertainment electronics) ... and whisper initiative into the ear of the user.«[40]

The difference between using and looking through, however, should not frighten us since it not only concerns electronic devices but any system of complex products and control circuits created by man. Regularly, even intellectually gifted adults can barely explain how water comes out of the faucet and power out of the outlet. Changing a tire on a car may still be possible, but repairing an engine? Instrumental intelligence and practical intelligence are not and have never been equal. How did Otl Aicher put it? »Gripping is what is grasped.« We now can add: »The surface of the user interface is its depth.« Some commentators and augurs of product culture consider the iPad the zenith of the Apple cosmos because the market at some point would (have to) be saturated with these i-devices. This is countered by the fact that this company in the meantime has a more global grip on the entertainment market than any other company and, in addition, that all offers of the music, film and information industries can be installed on the i-devices: from Microsoft and Google to Sony and Nintendo, and even to Warner Brothers and Disney. In this respect, the company represents the furthest reaching networking of communication forums ever. A change of these networking and convergence strategies cannot be expected for a long time because they are without an alternative in terms of economics, psycho-sociology, emotion and habit strategy.[41] But, above all, the Apple products are libidinously occupied and, in addition, more perfect in their design than their competitors. They are more intelligent in many ways: the i-worlds have been and will remain the real iQ-worlds of consumer electronics most likely for a long time to come.

259–262. David Jon Kassan: Painting with the app brushes on the iPad, 2010.
263. Kunstflug: Electronic finger calculator, 1986/87.

Anmerkungen

[1] Daniel Haas: »Ein Schatten seiner selbst«, in: *Frankfurter Allgemeine Zeitung*, 11. Dez. 2010, S. Z4 (Essay über Karl Lagerfeld).
[2] Frank Thadeusz: »Psychologie. Drang zum Ding. Experten rätseln über die bizarre sexuelle Spielart der Objektophilie«, in: *Der Spiegel*, Nr. 19, 2007, S. 160.
[3] Philipp Oehmke und Tobias Rapp: »Ein Jahrzehnt für die Ewigkeit (50 Jahre Beatles)«, in: *Stern*, Nr. 21, 2010, S. 109–121, S. 119 f.
[4] Roland Barthes: *Mythen des Alltags*, Frankfurt am Main 1964 (Paris 1957).
[5] Otl Aicher und Robert Kuhn: *Greifen und Griffe*, Köln 1987, darin: *Greifen und Begreifen*, S. 8 ff.
[6] Peter Sloterdijk: »Welt-Ortsgespräche«, in: Rudi Lamprecht: *Zukunft Mobile Kommunikation*, Frankfurt am Main 2001, S. 193–244, S. 201 f.
[7] Andererseits hat derzeit die Frage Konjunktur, wie es sich ohne Internet lebt. Vgl.: Alex Rühle: *Ohne Netz*, Stuttgart 2010; Christoph Koch: *Ich bin dann mal offline. Ein Selbstversuch. Leben ohne Internet und Handy*, München 2010; Susanne Beyer: »Ich bin dann mal off. Über die Kunst des Müßiggangs im digitalen Zeitalter«, in: *Der Spiegel*, Nr. 29, 2010, S. 56–67. Das Cover für die Titelstory zitierte J. H. W. Tischbeins *Bildnis Goethes in der römischen Campagna* (1786/87). Der zeitgenössische Protagonist hat nun allerdings ein Mobiltelephon am Ohr und eine Baseball-Cap auf dem Haupt. Begonnen hatte diese Diskussion der Internet-Askese Frank Schirrmacher mit seinem Buch *Payback*, München 2009.
[8] Neil Postman: *Wir amüsieren uns zu Tode. Urteilsbildung im Zeitalter der Unterhaltungsindustrie*, Frankfurt am Main 1986.
[9] Vgl. Volker Fischer: »Emotionen in der Digitale. Eine Phänomenologie elektronischer ›devices‹«, in: Bernhard E. Bürdek: *Der digitale Wahn*, edition suhrkamp, Frankfurt am Main 2001, S. 44–64.
[10] Vgl. Bruce Haring: *MP3. Die digitale Revolution in der Musikindustrie*, Freiburg 2002.
[11] Haring, a. a. O., S. 108–124: »Schuld war nur der Rio. Der erste tragbare MP3-Player.«
[12] Sloterdijk, a. a. O., S. 240 ff.
[13] Vgl. http://de.wikipedia.org/wiki/Apple.
[14] Zit. nach Christian Wurster: *Der Computer. Eine illustrierte Geschichte*, Köln 2002, darin: »Die Erfindung der graphischen Bedienoberfläche«, S. 228 ff.; vgl. auch Bernhard E. Bürdek: *Der Apple Macinthosh*, Frankfurt am Main 1997 (Reihe *Design-Klassiker*, Nr. 9, hrsg. von Volker Fischer).
[15] Vgl. Volker Fischer: »Apple Macinthosh Frogdesign Hartmut Esslinger«, in: Volker Albus et al.: *Design! Das 20. Jahrhundert,* München, London, und New York 2000, S. 160 f. Vgl. außerdem http://de.wikipedia.org/wiki/ Apple iMac.
[16] Vgl. http://de.wikipedia.org/wiki/iMac.
[17] Vgl. http://de.wikipedia.org/wiki/iPod.
[18] Zit. nach http://www.google.de/Andy Molloy: »Apple Computer Reading List, bn (= burn notice)«, *Network Forum*, 23. Okt. 2001.
[19] Michael Spehr: »Logitech Audio Station für den iPod. Die Günstige Docking Station«, in: *Frankfurter Allgemeine Zeitung*, 16. Jan. 2007.
[20] Thomas Wagner: »Berühre die Welt. Mit dem ultimativen Touch: Apples neues iPhone«, in: *Frankfurter Allgemeine Zeitung*, 11. Jan. 2007. Vgl. zum iPhone 4 Michael Spehr: »Schicke Schale und schöne Schnappschüsse«, in: *Frankfurter Allgemeine Zeitung*, 29. Jun. 2010.
[21] Zit. nach Thomas Wagner, a. a. O.
[22] Thomas Wagner, a. a. O.
[23] Vgl. Jochen Gros: »Neue Bilderschrift« sowie Volker Fischer: »Die diskursive Logik des Präsentativen. Anmerkungen zum Projekt »›Bilderschrift‹ von Jochen Gros«, in: Jochen Gros: *Digital Fiktional*, Frankfurt am Main 1993.
[24] Vgl. http://de.wikipedia.org/wiki/iPhone Outfits.
[25] Ulf Schönert: »Das neue Puschenkino«, in: *Stern*, Nr. 50, 2010, S. 83–90.
[26] Jordan Mejias: »Die Welt leuchtet auf. Der iPad ist da«, in: *Frankfurter Allgemeine Zeitung*, 6. Apr. 2010.
[27] Klaus Brinkbäumer, Thomas Schulz: »Der i-Kult. Wie Apple die Welt verführt«, in: *Der Spiegel*, Nr. 17, 2010, S. 66–78.
[28] Vgl. Dagmar Steffen: *Design als Produktsprache. Der »Offenbacher Ansatz« in Theorie und Praxis*, mit Beiträgen von Bernhard E. Bürdek, Volker Fischer, Jochen Gros, Frankfurt am Main 2000.
[29] Schönert, a. a. O., S. 88.
[30] Vgl. Isabell Hülsen und Martin U. Müller: »Angriff der Trittpadfahrer«, in: *Der Spiegel*, Nr. 26, 2010, S. 122; vgl. auch Michael Spehr: »Forscher Wettstreit der flachen Webpads«, in: *Frankfurter Allgemeine Sonntagszeitung*, 5. Sep. 2010, S. V10; vgl. auch Matthias Matting: »Die Tablet-Welle«, in: *Focus*, Nr. 35, 2010, S. 78 f.
[31] Vgl. http://de.wikipedia.org/wiki/Apple sowie Apple Inc.: *Report on iPod Manufacturing*, August 2006.
[32] Brinkbäumer, Schulz, a. a. O.
[33] Brinkbäumer, Schulz, a. a. O., S. 73.
[34] Markus Brauck, Isabell Hülsen und Marcel Rodenbach: »Steve sieht alles«, in: *Der Spiegel*, Nr. 26, 2010, S. 120–122; vgl. auch J. Hirzel, M. Franke, M. Kietzmann und A. Kusitzky: »Der große Imperator. Wie Steve Jobs mit Apple die Welt revolutioniert und bevormundet«, in: *Focus*, Nr. 35, 2010, S. 124–130.
[35] Vgl. Brauck et al., a. a. O., S. 120.
[36] Georg Meck: »An diesem Apfel misst sich die Welt«, in: *Frankfurter Allgemeine Sonntagszeitung*, 4 Apr. 2010, S. 31
[37] Sloterdijk, a. a. O., S. 209 f.
[38] Vgl. Volker Fischer: *Design heute. Maßstäbe. Formgebung zwischen Industrie und Kunststück*, München 1988, S. 288.
[39] Vgl. Matthias Weiß: »Mobiles Stillleben. Revolutioniert das Smartphone die Produktionsweise der Künstler?«, in: *Kunstzeitung*, Nr. 1, 2011, S. 1; vgl. auch die Ausstellung »David Hockney: Fleurs fraîches«, Fondation Bergé, Paris.
[40] Sloterdijk, a. a. O., S. 242 f.
[41] Zudem dreht sich das Produktkarussell der mobilen Internet-Geräte immer schneller. Pro Monat erscheinen 10 bis 15 Neuprodukte, oft allerdings für spezielle Märkte.

Notes

[1] Daniel Haas: »Ein Schatten seiner selbst«, in: *Frankfurter Allgemeine Zeitung*, 11 Dec. 2010, p. Z4 (essay about Karl Lagerfeld).

[2] Frank Thadeusz: »Psychologie. Drang zum Ding. Experten rätseln über die bizarre sexuelle Spielart der Objektophilie«, in: *Der Spiegel*, no. 19, 2007, S. 160.

[3] Philipp Oehmke und Tobias Rapp: »Ein Jahrzehnt für die Ewigkeit (50 Jahre Beatles)«, in: *Stern*, no. 21, 2010, S.109–121, S. 119 f.

[4] Roland Barthes: *Mythen des Alltags*, Frankfurt am Main 1964 (Paris 1957).

[5] Otl Aicher and Robert Kuhn: *Greifen und Griffe*, Köln 1987, darin: *Greifen und Begreifen*, pp. 8 ff.

[6] Peter Sloterdijk: »Welt-Ortsgespräche«, in: Rudi Lamprecht: *Zukunft Mobile Kommunikation*, Frankfurt am Main 2001, pp. 193–244, pp. 201 f.

[7] On the other hand, the question about what life would be like without the Internet is currently circulating. Cf. Alex Rühle: *Ohne Netz*, Stuttgart 2010; Christoph Koch: *Ich bin dann mal offline. Ein Selbstversuch. Leben ohne Internet und Handy*, München 2010; Susanne Beyer: »Ich bin dann mal off. Über die Kunst des Müßiggangs im digitalen Zeitalter«, in: *Der Spiegel*, no. 29, 2010, pp. 56–67. The cover for the cover story cited J. H. W. Tischbein's *Goethe in the Roman Campagna* (1786/87). The contemporary protagonist, however, now has a mobile phone at his ear and a baseball cap on his head. Frank Schirrmacher launched this discussion of Internet ascetics with his book *Payback*, Munich 2009.

[8] Neil Postman: *Wir amüsieren uns zu Tode. Urteilsbildung im Zeitalter der Unterhaltungsindustrie*, Frankfurt am Main 1986 (English original: *Amusing Ourselves to Death. Public Discourse in the Age of Show Business,* New York 1985).

[9] Cf. Volker Fischer: »Emotionen in der Digitale. Eine Phänomenologie elektronischer ›devices‹«, in: Bernhard E. Bürdek: *Der digitale Wahn*, edition suhrkamp, Frankfurt am Main 2001, pp. 44–64.

[10] Cf. Bruce Haring: *MP3. Die digitale Revolution in der Musikindustrie*, Freiburg 2002.

[11] Haring, loc.cit., pp.108–124: »Schuld war nur der Rio. Der erste tragbare MP3-Player.«

[12] Sloterdijk, loc. cit., pp. 240 ff.

[13] Cf. http://de.wikipedia.org/wiki/Apple.

[14] Quoted in Christian Wurster: *Der Computer. Eine illustrierte Geschichte*, Cologne 2002, therein: »Die Erfindung der graphischen Bedienoberfläche«, pp. 228 ff; also cf. Bernhard E. Bürdek: *Der Apple Macinthosh*, Frankfurt am Main 1997 (series *Design Classics*, no. 9, ed. by Volker Fischer).

[15] Cf. Volker Fischer: »Apple Macintosh Frogdesign Hartmut Esslinger«, in: Volker Albus et al.: *Design! Das 20. Jahrhundert*, Munich, London, and New York 2000, pp.160 ff. Also cf. http://de.wikipedia.org/wiki/ Apple iMac.

[16] Cf. http://de.wikipedia.org/wiki/iMac.

[17] Cf. http://de.wikipedia.org/wiki/iPod.

[18] Quoted in http://www.google.de/Andy Molloy: »Apple Computer Reading List, bn (= burn notice)«, *Network Forum*, 23 Oct. 2001.

[19] Michael Spehr: »Logitech Audio Station für den iPod. Die günstige Docking-Station«, in: *Frankfurter Allgemeine Zeitung*, 16 Jan. 2007.

[20] Thomas Wagner: »Berühre die Welt. Mit dem ultimativen Touch: Apples neues iPhone«, in: *Frankfurter Allgemeine Zeitung*, 11 Jan. 2007. Cf. on the iPhone 4, Michael Spehr: »Schicke Schale und schöne Schnappschüsse«, in: *Frankfurter Allgemeine Zeitung*, 29 Jun. 2010.

[21] Quoted in Thomas Wagner, loc. cit.

[22] Thomas Wagner, loc. cit.

[23] Cf. Jochen Gros: »Neue Bilderschrift« and Volker Fischer: »Die diskursive Logik des Präsentativen, Anmerkungen zum Projekt ›Bilderschrift‹ von Jochen Gros«, in Jochen Gros: *Digital Fiktional*, Frankfurt am Main 1993.

[24] Cf. http://de.wikipedia.org/wiki/iPhone Outfits.

[25] Ulf Schönert: »Das neue Puschenkino«, in: *Stern*, no. 50, 2010, pp. 83–90.

[26] Jordan Mejias: »Die Welt leuchtet auf. Der iPad ist da«, in: *Frankfurter Allgemeine Zeitung*, 6 Apr. 2010.

[27] Klaus Brinkbäumer and Thomas Schulz: »Der iKult. Wie Apple die Welt verführt«, in: *Der Spiegel*, no. 17, 2010, pp. 66–78.

[28] Cf. Dagmar Steffen: *Design als Produktsprache. Der »Offenbacher Ansatz« in Theorie und Praxis*, with contributions by Bernhard E. Bürdek, Volker Fischer and Jochen Gros, Frankfurt am Main 2000.

[29] Schönert, loc. cit., p. 88.

[30] Cf. Isabell Hülsen, Martin U. Müller: »Angriff der Trittpadfahrer«, in: *Der Spiegel*, no. 26, 2010, p. 122. Cf. also: Michael Spehr: *Forscher Wettstreit der flachen Webpads*, in: *Frankfurter Allgemeine Sonntagszeitung*, 5 Sep. 2010, p. V10; see also: Matthias Matting: »Die Tablet-Welle«, in: *Focus*, no. 35, 2010, pp. 78 f.

[31] Cf. http://de.wikipedia.org/wiki/Apple and Apple Inc: *Report on iPod Manufacturing*, Aug. 2006.

[32] Brinkbäumer, Schulz, loc. cit. p. 72.

[33] Brinkbäumer, Schulz, loc. cit. p. 73.

[34] Markus Brauck, Isabell Hülsen, Marcel Rodenbach: »Steve sieht alles«, in: *Der Spiegel*, no. 26, 2010, pp. 120–122; see also: J. Hirzel, M. Franke, M. Kietzmann and A. Kusitzky: »Der große Imperator. Wie Steve Jobs mit Apple die Welt revolutioniert und bevormundet«, in: *Focus*, no. 35, 2010, pp. 124–130.

[35] Cf. Brauck et al., loc. cit., p. 120.

[36] Georg Meck: »An diesem Apfel misst sich die Welt«, in: *Frankfurter Allgemeine Sonntagszeitung*, 4 Apr. 2010, p. 31.

[37] Sloterdijk, loc. cit., p. 209 ff.

[38] Cf. Volker Fischer: *Design heute. Maßstäbe. Formgebung zwischen Industrie und Kunst-Stück*, Munich 1988.

[39] Cf. Matthias Weiß: »Mobiles Stillleben. Revolutioniert das Smartphone die Produktionsweise der Künstler?«, in: *Kunstzeitung*, no. 1, 2011, p. 1. Cf. also the exhibition »David Hockney: Fleurs fraîches«, Fondation Bergé, Paris.

[40] Sloterdijk, loc. cit., p. 242 f.

[41] In addition, the carousel of mobile Internet products is spinning at an increasing speed. 10 to 15 new products are launched each month, but often for special markets.

Nutzer-Erfahrungen

Produkte werden geplant, entworfen, distribuiert und dann in den Markt gebracht. Die Käufer erwerben das Produkt und gehen als Nutzer fürderhin mit ihm um. So werden sie zu Vollstreckern und Vollendern, Sachwaltern und Scouts all der ursprünglich intendierten Intentionen. Bei den i-Produkten, die sich ausdrücklich über Nutzerführungen (-verführungen?) definieren, wäre eine Nichtberücksichtigung von Nutzer-Erfahrungen eine gravierende Unterlassung. Man kann Produkte, ihre designhistorische und gesellschaftliche Position analysieren, man kann Benutzeroberflächen und Nutzungen beschreiben und interpretieren, aber, um mit Norbert Elias zu sprechen, neben der »Soziogenese« steht gleichberechtigt die »Psychogenese«. Diese erdet die zwangsweise abstrahierenden Analysen der historischen und sozialen Stellung von Produkten in Erfahrungsberichten des je individuellen Gebrauchs. Es geht um Innenansichten auf Außenansichten. Neben der inzwischen etablierten »oral history« sollte es auch eine »doing history« für Gebrauchsgegenstände geben. Die folgenden Erfahrungsberichte sind dafür ein Plädoyer.

Janine, Gymnasiastin, 16

Ich bin heute 16 Jahre alt. Mit 12 bekam ich meinen ersten iPod. Das war noch der heute iPod classic genannte Player, heute habe ich einen türkisen iPod nano. Seit Jahren benutze ich ihn auf dem Schulweg, aber auch manchmal im Unterricht oder zu Hause bei den Hausaufgaben. Mit 14 hat mir mein Vater den iPod abgenommen, weil ich so viel gehört habe, daß ich nichts mehr mitbekommen habe und sitzengeblieben bin. Ich lade mir hauptsächlich indie pop, dance punk und fantasy pop, aber auch AC/DC herunter und höre die Songs dann in zufälliger Reihenfolge (kann man einstellen). Einmal ist mir in der U-Bahn mein iPod geklaut worden, aber ich habe auf ein neues gespart und meine Mutter hat mir 80 Euro dazugegeben, was mein Vater nicht wissen durfte. Mein älterer Bruder nennt mich immer iPodine, was mich manchmal nervt, aber nicht wirklich stört. Zu meinem nächsten Geburtstag wünsche ich mir ein neues iPhone, wo der iPod schon drin ist.

Marc, Tänzer, 21

Aus beruflichen Gründen ziehe ich das streichholzschachtelgroße iPod shuffle vor. Ich kann es an jedes Trikot anclipsen und benutze es oft bei Proben, die ich allein, oft zu Hause, mache, bevor die Proben mit der Tanzcompagnie stattfinden. Da ist mir der vorgegebene Rhythmus im Ohr wichtiger als ein perfekter Klang. Denn da geht es um die Einübung von Tonrhythmen und Bewegungen. Beides muß sich im Unterbewußtsein verbinden. In dem modernen Ballett *Jeu de carte* z. B. ist in der Inszenierung, die ich meine, der Bühnenraum nur mit UV-Licht beleuchtet, also im Grunde dunkel. Zwölf schwarz gekleidete und geschminkte Tänzer tanzen jeweils mit einem Laserschwert. Man sieht eigentlich als Zuschauer nur die schwebenden, reflektierenden Klingen. Da ist absolute Präzision gefragt, etwa so, wie bei den *Backstage*-Tänzern in vielen Videos von Michael Jackson.

Holger, Abiturient, 19

Ich habe mir zum Abitur von meinen Eltern ein iPad gewünscht. Ein iPhone hatte ich schon zwei Jahre vorher. Ich habe es auch tatsächlich bekommen. Klar, daß ich es auf die Abi-Feier mitgenommen habe. Die Freundinnen mancher meiner Kumpels waren ein bißchen sauer, weil die, statt sich mit ihnen zu beschäftigen, alle um mein iPad herumstanden. Jeder wollte es mal ausprobieren und daran herumfingern. Nach dem Prüfungstrubel bin ich dann erst mal an die Costa del Sol gefahren, natürlich mit dem iPad, und manche *Apps* haben mir auch dort unten gute Dienste geleistet: Banking, Hotels, Shopping, Stadtpläne, Dates über *YouTube*…. Das iPad wird mein »ständiger Begleiter« bleiben, wenn meine Freundin dies toleriert.

Michael, Berufssportler, 24

Ich hatte schon mit 12 einen *Walkman* von Sony, dann einen *Discman* von Panasonic. Insofern bin ich seit mehr als zehn Jahren an Musik beim Joggen und beim Training gewöhnt. Ich möchte das nicht missen. Seit fünf Jahren habe ich einen iPod nano, auf den ich sehr gezielt auch Songs überspiele, die mir beim Training rhythmisch helfen. Der einzige Nachteil ist, wenn ich in einer Gruppe trainiere, daß man sich nichts zurufen kann.

Roland, Komponist, 45

Ich habe eine sehr individuelle Nutzung des iPods für mich entwickelt. Aus vielen, vielen Kompositionen im Netz, die ich über das iPod hole, sample ich etwas Eigenes, indem ich es in meinem *Powerbook* zusammenstelle, wie man auch sagt, eine Bricollage oder ein Mash up. Um den Raumklang zu überprüfen, kommt diese dann zurück auf den iPod und von dort auf meine »Docking Station« von Bose. Bei vollem Raumklang diskutiere ich dann mit Freunden meine Kompositionen. Mein iPod ist also definitiv ein Arbeitstool.

Bernd, Fliesenleger, 38

Kurz vor der Fußball-Weltmeisterschaft 2010 bin ich schwer erkrankt und mußte ins Krankenhaus. Es war ein Dreibettzimmer mit lauter bettlägerigen Schwerkranken, Fernsehen nicht erlaubt. Meine Frau hat mir den iPod classic ins Krankenzimmer geschmuggelt, und ich habe alle WM-Spiele (wie damals als Junge die Karl-May-Romane) unter der Bettdecke geguckt. So war ich immer auf dem laufenden, aber bei der Nachmittagsvisite mußte ich manchmal vorsichtshalber vorher die Kiste ausmachen. Die Abendspiele waren ja glücklicherweise lange nach der Ausgabe des Nachtessens. Das einzig Schlechte war, daß ich für unsere Jungs nicht jubeln konnte. Ich glaube, daß mir der iPod beim Gesundwerden schon ein bißchen geholfen hat.

Eva, Kulturwissenschaftlerin, 51

Den folgenden Text übermittle ich über mein iPhone. Ich liebe mein iPhone und will die anderen Mobiltelephone nicht mehr benutzen. Es war keine Liebe auf den ersten Blick, keine Beziehung aus nur optischen Gründen. Vielmehr handelt es sich um eine Leidenschaft, die sich beim Entde-

User experiences

Products are planned, designed, distributed, and then launched on the market. People purchase products and from then on deal with them as users. They thus become the executors and terminators, advocates and scouts of all of the originally envisioned intentions. In the case of i-products that expressly define themselves via the user guidances (temptations?), the ignorance of user experiences would be a striking omission. Products, their design historic and social position, can be analyzed, user interfaces and uses can be described and interpreted; however, to paraphrase Norbert Elias, the »psychogenesis« stands right next to and equal to the »sociogenesis.« It necessarily grounds the abstracting analyses of the historic and social position of products in field reports of the pertaining individual use. It is about internal views on external views. In addition to the meanwhile established »oral history« there should also be a »doing history« for objects of daily use. The following reports about personal experiences are a plea for the latter.

Janine, high school student, 16
I'm 16 years old. I got my first iPod when I was twelve. It was the one that is called iPod classic today; now I have a turquoise iPod nano. I have been using it on my way to school for years, but sometimes also during lessons or at home when I am doing my homework. When I was 14 my father took the iPod away from me because I was listening to it so much that I lost track of what I was doing and had to repeat a class. I mainly download indie pop, dance punk, and fantasy pop, but also AC/DC, and then I listen to the songs randomly (an iPod setting). Once, my iPod was stolen in the subway, but I saved money to get a new one, and my mother chipped in 80 euro, which my father was not allowed to know. My older brother always calls me iPodine, which gets on my nerves sometimes but doesn't really bother me. For my next birthday I want to get an iPhone with an integrated iPod.

Marc, dancer, 21
For professional reasons I prefer the matchbox size iPod shuffle. I can clip it onto any jersey and often use it during practice sessions, which I do on my own, often at home, before practice with the dance company takes place. The rhythm in my ear is more important to me than perfect sound reproduction because this is about rhythm and movement. Both have to connect in the subconscious. In the modern ballet *Jeu de carte*, for example, the staging I have in mind is only lit with UV light, basically dark. Twelve dancers dressed in black and wearing black make-up each dance with a laser sword. As a viewer you really only see the floating, reflecting blades. This requires absolute precision, comparable to that of the background dancers in many of Michael Jackson's videos.

Holger, high school graduate, 19
I wanted to get an iPad from my parents for graduation. I had an iPhone two years earlier. And I actually got it. It was clear that I was going to take it to the graduation party. The girlfriends of some of my buddies were a bit upset because instead of taking care of them they all stood around my iPad. Everybody wanted to try it out and put their fingers on it. After the stress of final exams I first drove to the Costa del Sol, of course with my iPad, and several apps were very helpful: banking, hotels, shopping, city maps, dates via YouTube … The iPad will continue to be my »permanent companion« assuming my girlfriend tolerates it.

Michael, professional athlete, 24
I had a Sony walkman at the age of 12, then a Panasonic discman. In this regard, I have been accustomed to listening to music while jogging or training for more than ten years. I wouldn't want to be without it. I've owned an iPod nano for five years now and use it specifically for downloading songs that help me establish a rhythm while training. The only disadvantage is that when you train in a group, you can't communicate with others.

Roland, composer, 45
I have developed a very personal way of using my iPod. From many, many tracks online, which I get via the iPod, I sample something of my own by compiling it on my Powerbook, like a bricollage or mash-up. To check the spatial sound, I then put it back on the iPod and from there to my Bose docking station. When I've achieved full spatial sound I discuss my compositions with my friends. My iPod is hence definitely a working tool.

Bernd, tile layer, 38
Shortly before the FIFA World Cup 2010 I became seriously ill and had to go to hospital. It was a room with three beds, and all of us were seriously ill, television was not allowed. My wife smuggled the iPod classic into the hospital and I was able to watch all of the World Cup matches (like I was able to read the Karl May books when I was a young boy) under the blanket. I was always up-to-date, but when the afternoon visit from the doctor was approaching, I sometimes had to switch off the device as a precaution. The evening matches fortunately took place long after dinner was served. The only bad thing was that I couldn't cheer for our boys. I believe that the iPod did help me a bit with my recovery.

Eva, cultural scientist, 51
I am sending this text via my iPhone. I love my iPhone and don't even want to use the other mobile phones anymore. It wasn't love at first sight, no relationship for visual reasons. It is rather a passion that was raised while discovering that everything can be handled without a problem and of all the things that can be simultaneously combined. When writing SMS I no longer needed to count when typing the letters and hence there are no more errors instead, there's a comfortably sliding keyboard. The address, already stored in the contacts list, does not have to be re-entered but is automatically displayed. I listen to music when I'm in the subway, and when the phone rings the

cken, daß alles mühelos handhabbar ist und was alles gleichzeitig miteinander kombinierbar ist, einstellte. Beim SMS-Schreiben fielen das Zählen und damit das Verzählen beim Eintippen der Buchstaben weg. Statt dessen eine bequem gleitende Tastatur. Die Adresse, schon unter Kontakte gespeichert, muß nicht eigens wieder eingegeben werden, sondern blendet sich automatisch ein. Ich höre iPod in der U-Bahn, es klingelt, die Musik wird leise, ich nehme das Gespräch an, und nach Beenden ist sie mit einem kurzen Crescendo wieder hörbar. In Portugal suche ich via iPhone das Meer. Ich klicke auf »Karten« und gebe als Ziel den Strand ein. Dort angekommen bin ich von den rost-orangen Felsen und den türkisblauen Wellen begeistert. Sogleich mache ich Photos und sende sie vom Badetuch aus als E-Mails – das alles mit dem iPhone – an meinen Mann. Die Rechnung dafür kam später und war hoch, sehr hoch, aber immer noch kein Grund, mir meine naive Begeisterung über eine Realität gewordene Utopie von Technik aus Unkompliziertheit und Omnipotenz nehmen zu lassen. Eine sehr neue Erfahrung.

Marianne, Oberärztin, 43

Seit längerem schon sind Tablet-Computer bei Visiten in Krankenhäusern als Diagnose-Hilfsgeräte im Einsatz. Ich bin also eigentlich schon vor dem iPad damit vertraut gewesen. Ich habe mir dann auch ein iPad gekauft, weil ich es auch privat nutze. Beruflich rufe ich damit medizinische Vergleichsdaten ab, von unserer Klinik, aber auch von anderen Krankenhäusern. Ich speichere die Dienstpläne, Vertretungen, Urlaubszeiten und Krankenausfälle und bin so immer in der Lage, auf dem aktuellen Stand des verfügbaren Personals Einsatzentscheidungen zu treffen. Zu Hause muß ich das iPad manchmal mit Zähnen und Klauen verteidigen, denn meine zehn- und zwölfjährigen Kinder finden es inzwischen besser als das Fernsehen.

Jörn, Kommunikationsdesigner, 20

Ich bin mit dem iPod, später mit dem iPhone groß geworden. Als Designer habe ich natürlich immer Apple bevorzugt, ob Macs oder PowerBooks. Das braucht man, glaube ich, nicht eigens zu erklären. Als ich die ersten Infos über das iPad bekam und las, wollte ich es sofort haben. Ich habe mich eine ganze Nacht vor dem Apple Store angestellt und wow! Endlich um 8:30 Uhr hatte ich das Ding ergattert. Es ist ja nicht wirklich ein Profigerät, aber in der Kombination von mobilem Internet, Apps, Büroanwendungen, Videos, Nachrichten, Spielen und Zeitschriften einfach geil. Ich nutze das iPad beruflich und privat und habe überhaupt nichts dagegen, wenn beides sich ergänzt. Jedenfalls ist mir dieses Wunderbrett ans Herz gewachsen; – als Ergänzung wohlgemerkt zu meinem PowerBook und meinem neuen iMac, den ich mir 2009 gekauft habe.

Yun Jin, Programmierer, 25

Während eines Stipendiums in den USA habe ich mir ein iPhone gekauft. Ich mußte drei Jahre darauf sparen, aber ich finde, es hat sich gelohnt. Zurück in Shanghai, dachte ich, ich wäre einer von Wenigen mit einem solchen Gerät. Weit gefehlt! Fast jeder meiner Freunde in Shanghai hat ein iPhone oder ein Google-Smartphone. Zwei Drittel davon sind allerdings Fälschungen, funktionieren aber trotzdem. Meine Freundin Yao hat ein Smartphone, wo Gooble draufsteht und mein Kumpel Zong hat vor kurzem auf einem Flohmarkt in Shanghai ein Mobiltelephon erworben mit dem Namen iFabulous. Überhaupt sind wir ja von Produkt-Klonen umgeben. Auf meiner Sporttasche steht Adadis statt Adidas und Sony heißt bei uns Sqny.

Katharina, Modedesignerin, 32

Aufgrund vieler Location-Termine bin ich beruflich auf ein Handy angewiesen. Ich habe eigentlich alle Anbieter durch, Nokia, Siemens, Samsung, LG, Motorola. Heute habe ich ein iPhone, nicht nur wegen seiner Eleganz, sondern auch wegen seiner Programme. Ich benutze es zwar häufig zum Dating, aber ebenso häufig zum Musikhören oder um Nachrichten anzuschauen. Das alles kann ich nebenbei und überall, und das heißt spontan machen. Vor allem finde ich die Touch-Funktion toll und daß man keine Tasten mehr braucht.

Marian, Banker, 33

Mein iPhone habe ich gleich nach seinem Erscheinen auf einer Geschäftsreise 2007 in den USA gekauft, natürlich ohne Vertrag. In Deutschland habe ich mir dann Programme und Anwendungen aus dem Netz geholt: Das war viel billiger als ein Vertrag mit T-Mobile. Ich gehöre halt zu diesen Leuten, die neue Tools immer als erster haben müssen. Das bringt mir auch Aufmerksamkeit in meinem Freundeskreis. Deshalb habe ich mich auch geärgert, daß es bei der Auslieferung des iPad einen Engpaß gab.

Die protokollierten Nutzer-Erfahrungen belegen eine umfangreiche Neugier gegenüber den Möglichkeiten der i-Geräte. Sie regen zu interaktiven Versuchen und Vereinnahmungen an und »flüstern«, mit einer schönen Formulierung von Sloterdijk, »dem Benutzer Initiative ein«. Die Besitzer der Geräte kreieren mit ihrem Nutzerverhalten sozusagen ihre je individuellen Apps. So wie Tageszeitungen mittlerweile Leserphotos von privaten »Bildreportern« veröffentlichen und damit das Informationsangebot demonopolisieren, so individualisieren die Nutzer durch die Art und Weise ihres Gebrauchs die Geräte handlungsstrategisch. Die Trennung der klassischen Rhetorik zwischen »überzeugen« und »überreden« überwinden die i-Geräte bravourös. In der Tat: die Psyche zu »überwinden« beschreibt vielleicht am besten den Anmutungscharakter dieser Gerätefamilie, der eine »digitale Sinnlichkeit« zu eigen ist.

volume of the music decreases, I accept the call and after finishing it the music is again audible after a short crescendo. In Portugal I search for the sea via iPhone. I click on »maps« and enter the beach as my destination. Having arrived there I am thrilled with the rusty orange rocks and turquoise waves. I immediately take pictures and e-mail them to my husband while lying on my bath towel – everything with the iPhone. The bill for it came later and was high, very high, but still no reason for me to allow anyone to take away my naïve enthusiasm about a utopia of technology consisting of simplicity and omnipotence that has become a reality – a very new experience.

Marianne, senior physician, 43

Tablet computers have long been used for ward rounds in hospitals as a diagnosis assistant tool. So I was familiar with them even before the iPad. I then went and bought an iPad because I also use it in private. At work I access medical comparative data from our hospital but also from others. I store the duty schedules, stand-ins, holiday schedules and absences and I am always up-to-date with the available staff for making assignment decisions. At home, I sometimes have to defend the iPad tooth and nail because my ten and twelve year-old children like it more than television.

Jörn, communication designer, 20

I grew up with the iPod, later with the iPhone. As a designer I naturally always preferred Apple, whether Macs or PowerBooks. I don't think this requires explanation. When I received the first information about the iPad and read it, I wanted to get one right away. I stood in line outside the Apple store for a whole night and – wow! – by 8:30 in the morning I had one. It's not really a professional device, but in combination with mobile Internet, apps, office applications, videos, news, games and magazines it's simply awesome. I use the iPad at work and in private and I don't mind if the two overlap a bit. Anyway, this magical tablet has really found a place in my heart – as an addition, mind you, to my PowerBook and my new iMac, which I bought in 2009.

Yun Jin, programmer, 25

I bought an iPhone during my scholarship in the USA. I had to save for it for three years, but I think that was well worth it. Back in Shanghai I thought I was one of few with such a device. Far from it! Almost all of my friends in Shanghai had an iPhone or a *Google* smartphone. Two thirds of them, however, are counterfeits, but they work anyway. My girlfriend Yao has a smartphone with the logo *Gooble* on it, and my buddy Zong recently bought a mobile phone at a flea market in Shanghai named *iFabulous*. Anyway, we're surrounded by product clones. My sports bag displays *Adadis* instead of *Adidas*, and *Sony* is *Sqny*.

Katharina, fashion designer, 32

Due to many location dates I need to have a mobile phone for professional reasons. I've basically gone through all of the suppliers – Nokia, Siemens, Samsung, LG, Motorola. Today I have an iPhone, not only because of its elegance but also because of its programs. I often use it for dating but also often for listening to music or watching news. I can do all of that on the side and anywhere, and that means spontaneously. Above all, the touch-sensitive screen is great, and the fact that you no longer need keys.

Marian, banker, 33

I bought my iPhone just as it was being launched during a business trip to the USA in 2007, of course without a contract. When I returned to Germany, I downloaded programs and applications from the Internet: that was much cheaper than a contract with T-Mobile. I'm one of those people who have to be the first to own new tools. It also gets me attention among my friends. Therefore, I was upset when there was a bottleneck with the delivery of the iPad.

These recorded user experiences prove a comprehensive curiosity towards the possibilities of the i-devices. They stimulate interactive tests and appropriations and »whisper« – to use a beautiful formulation by Sloterdijk – »initiative into the ears of the users.« The owners of the devices create, so to speak, their individual *apps* through their user behavior. Like daily newspapers meanwhile publish reader's photos by private »photo reporters« and thus de-monopolize the information offer, the users individualize the devices in an action strategic way through the way of their use. The i-devices masterfully overcome the separation of classic rhetoric between »convincing« and »persuading«. In fact, »overcoming« the psyche perhaps best describes the image character of this family of devices, which has embraced a »digital sensuality«.

Glossar

Air Print
App für Drucken per Funk für iPhone und iPad

Android
Betriebssystem für Smartphones von Google

Apple Store
weltweit vertretene Geschäfte für Apple-Produkte

App Sharing
App-Übermittlung von einem Smartphone zu einem weiteren

App Sharing Widget
App-Übermittlung über Twitter, SMS oder E-Mails

Apps (Applications)
Anwendungsprogramme, teilweise kostenpflichtig

App Store
virtuelle Verkaufsplattform für Anwendungsprogramme

Augmented Reality (AR)
Ergänzung von Bildern oder Videos mit computergenerierten Zusatzinformationen

Beeper
mobiles Kommunikationsfunkgerät auf der Basis von Funk-Rufdiensten

BlackBerry®
Mobiltelephon mit zusätzlichen Organizer-Funktionen

Bluetooth
Schnittstelle für die Kommkunikation mobiler Kleingeräte untereinander und mit Computern und Peipheriegeräten

Bookmarks
Lesezeichen für häufig aufgesuchte Webseiten

Chat room
virtueller »Begegnungsraum« im Internet für »real time«-Unterhaltungen

Click Wheel
Rad-Wipptaste des iPods

Cover Flow
dreidimensionales Blättern durch Albencover auf dem iPod

Cross marketing
gegenseitige Werbung für Konkurrenz- bzw. Ergänzungsprodukte

Customization
Anpassung von Großserienprodukten an individuelle Nutzerprofile

Discman
mobiles Gerät für CDs (ab 1982)

Docking Station
Empfangsgeräte mit externen Lautsprechern für iPod, iPhone oder iPad

E-Book
elektronisches Gerät zum Lesen von Texten

eMate
kleines Schul-Notebook von Apple (in den 1990er Jahren nur in den USA erhältlich)

E-Reading
Lesen von elektronischen Texten, erweiterbar durch Videos

Facebook
Website zur Bildung und Unterhaltung sozialer Netzwerke

Facetime
Videotelephonate von iPhone zu iPhone

FireWire
Schnittstelle für schnellen Datenaustausch zwischen Computern und Multimedia- oder anderen Peripheriegeräten, von Apple entwickelt (seit 1995)

Flatrate
Pauschaltarif für Kommunikationsdienstleistungen

Flickr
Website für private Photographien im Internet

Freenet
dezentralisierte Tauschbörse für Musikdateien im Internet

FriendStream
zentralisierter Platz für alle gespeicherten sozialen Netzwerke

Gameboy
mobiles Gerät für Videospiele

Gigabyte
1000 Megabyte, also eine Milliarde Byte, 10^9, 1000 000 000 Byte (griech: gigas = Gigant)

GPS-Modul
»Global Positioning System«, weltweite Standortbestimmung durch Satelliten

Handhelds
auch PDAs genannt, z. B. Palm Pilot oder Newton

iAd
kostenpflichtiges App für Werbung auf den i-Geräten

iBook
Einsteiger-Notebook von Apple (1999–2006)

iLife
Multimedia-Paket von Apple für Bild-, Audio- und Videobearbeitung

Glossary

Air Print
App for printing via wireless for iPhone and iPad

Android
Operating system for Google smart phones

Apple Store
Stores for Apple products worldwide

App Sharing
Transfer of apps from one smart phone to another

App Sharing Widget
App transfer via Twitter, SMS or e-mail

Apps (Applications)
Application programs, some for a fee

App Store
Virtual sales platform for applications

Augmented Reality (AR)
Augmentation of images or videos with additional computer-generated information

Beeper
Small mobile device based on radio call services

BlackBerry®
Mobile phone with additional organizer functions

Bluetooth
Wireless technology for communication between small mobile devices and with computers and peripheral devices

Chat room
Virtual »meeting room« on the Internet for real-time chatting

Click Wheel
Control button on the iPod

Cover Flow
Three-dimensional leafing through album covers on the iPod

Cross marketing
Mutual advertising for competing or supplementary products

Customization
Adjustment of large serial products to individual user profiles

Discman
Mobile device for playing CDs (starting 1982)

Docking station
Device with speakers for iPod, iPhone or iPad

E-Book
A digital book readable on a computer or other digital device

eMate
Small school laptop by Apple (available only in the USA in the 1990s)

E-Reading
Reading of electronic texts, expandable by videos

Facebook
A social networking website

Facetime
Video telephone calls from iPhone to iPhone

FireWire
Interface for fast data exchange between computers and multimedia or other peripheral devices, developed by Apple (since 1995)

Flatrate
Flat fee for communication services

Flickr
Website for sharing photographs on the Internet

Freenet
Decentralized exchange for music files on the Internet

FriendStream
Smartphone app for displaying feeds from various social networks on a single page

Gameboy
Mobile device for video games

Gigabyte
1000 Megabyte, thus one billion byte, 10^9, 1 000 000 000 Byte (Greek: gigas = giant)

GPS-Modul
The Global Positioning System is a satellite-based navigation system

Handhelds
e.g., Palm Pilot or Newton

iAd
For-fee app for advertisements on i-devices

iBook
Lower-end laptop for consumer and education markets by Apple (1999–2006)

iLife
A suite of software applications developed by Apple for organizing, editing, and publishing photos, movies, and music

iMac
Personal computer by Apple (since 1984)

Image transfer
Strategic use of the reputation of a product or company for other products or companies

iMac
Personal Computer von Apple (seit 1984)

Image Transfer
strategische Nutzung des Renomees eines Produktes oder eines Unternehmens für andere Produkte oder Unternehmen

iMovie
Videoschnittprogramm von Apple für das MAC OS-Betriebssystem und das iPhone der 4. Generation

Internet-Browser
Programm zur Darstellung von Webseiten. Browser sind also die Benutzeroberflächen für Webanwendungen.

iPad
Tablet-Computer von Apple (seit 2010)

iPhone
das erste Mobiltelephon mit berührungssensitivem Display (seit 2007)

iPhone OS
Betriebssystem von Apple für Smartphones

iPhoto
Bildverwaltung von Apple

iPod
MP3-Player von Apple in verschiedenen Varianten und Gigabyte-Kapazitäten (seit 2001)

iPod touch
MP3-Player in der Form des iPhones

iTalk
Ton- und Sprachaufnahmen für i-Geräte

iTime
Macintosh-Systemerweiterung durch eine atomgenaue Uhr aus dem Internet

iTrip
Konvertierung der Audiosignale des iPods in UKW-Radiosignale für den Empfang in Autoradios

iTunes
Apple-Software für das Downloaden und Verwalten von Musikstücken

Konvergenz
Ineinanderübergehen verschiedener Geräte bzw. Funktionen

MacBook
kleines Notebook

MacBook Pro
Profigerät mit bis zu zwölf Prozessorkernen

Magic Mouse
Computermaus von Apple mit Multitouch-Technologie

Magic Trackpad
Nachfolgegerät für Computermaus von Apple, das über Bluetooth mit den iMac-, MacBook Pro- und Mac mini-Computern kommuniziert

Megabyte
eine Million Byte, 10^6, 1 000 000 Byte (griech: megas = groß)

Minidisc
kleinere Variante von Audio-CDs

MobileMe
kostenpflichtige App für E-Mails, Photographien, Adressen und Termine für Apple-Geräte

MP3-Datei
Kompressionsstandard zur Verdichtung von Musikdateien zum Downloaden im Internet

MP4-Datei
Kompressionsstandard zur Verdichtung von Film- und Musikdateien zum Downloaden im Internet

Multitasking
gleichzeitiges Aufrufen mehrerer Apps

MySpace
Website zur Bildung und Unterhaltung sozialer Netzwerke

Napster
kostenloser Servive zum Finden von Musiktiteln im Internet über das Netzwerk aller Nutzer, aber über zentrale Server, also eine Kombination von Chatten, Musik-Player und Suchfunktion (Frühjahr 1999 bis Juli 2001, dann durch Gerichtsurteil untersagt)

Nerds
selbstironische Bezeichnung in Internet-Communities für Computerfreaks

Netbook
kleines mobiles Notebook

Netscape
erste kommerzielle Internetfirma (1993 gegründet, seit 1995 an der Börse, 1999 Übernahme durch AOL)

Newton
Message Pad mit Handschriftenerkennung

Notebook
tragbarer PC (ab 1982)

Open-Source-Programs
frei zugängliche Software-Programme

Organizer
miniaturisierte Taschenrechner mit programmierbaren Funktionen

Pager
mobiles Kommunikationsgerät auf der Basis von Funk-Rufdiensten

Palm Pilot
Taschencomputer mit Betriebssystem, Rechner, Adreßbuch und Datenbank (der erfolgreichste PDA der 1990er Jahre, der einen eigenen Standard begründete)

iMovie
Video editing software application from Apple for the MAC OS operating system and the 4th generation iPhone

Internet browser
Software application for displaying information resources on the World Wide Web. Browsers are user interfaces for web applications

iPad
Tablet computer by Apple (since 2010)

iPhone
The first mobile phone with a touch-sensitive display

iPhone OS
Operating system from Apple for smart and touch phones

iPhoto
Software application from Apple for editing and organizing digital photos

iPod
MP3 player from Apple in various versions and storage capacities

iPod touch
MP3 player in the design of the iPhone

iTalk
Audio and language recordings for i-devices

iTime
Macintosh system expansion via an atomic clock from the Internet

iTrip
Conversion of the audio signals of the iPod into VHF radio signals for reception on car radios

iTunes
Apple software for organizing, playing and downloading music tracks

Convergence
Convergence of different devices or functions

MacBook
Small notebook

MacBook Pro
High-end laptop with up to twelve processor cores

Magic Mouse
Computer mouse by Apple with multi-touch technology

Magic Trackpad
Successor device of computer mouse from Apple, which communicates with iMac, MacBook Pro and Mac mini computers via Bluetooth

Megabyte
One million bytes, 10^6, 1 000 000 bytes (Greek: megas = large)

Minidisc
Smaller version of audio CDs

MobileMe
A subscription-based suite of online services and software applications for e-mails, photographs, addresses, and appointments on Apple devices

MP3 files
A digital audio encoding format for the compression of music files

MP4 files
A digital audio encoding format for the compression of video and music files

Multitasking
Simultaneous use of several apps

MySpace
A social networking website

Napster
Free service for finding music tracks on the Internet via a network of all users, but via central servers – i.e., a combination of chat, music player and search (spring 1999 to July 2001, then shut down via court decision)

Nerds
Selfironic term used in online communities for a computer freak

Netbook
Small mobile notebook

Netscape
First commercial Internet company, established in 1993, since 1995 publicly listed, taken over by AOL in 1999

Newton
Message pad with handwriting recognition

Notebook
Portable PC (starting 1982)

Open-source programs
Software programm license-free

Organizer
Miniaturized pocket calculator with programmable functions

Pager
Mobile communication device based on radio call services

Palm Pilot
Pocket computer with operating system, calculator, address book, and database (the most successful PDA in the 1990s, which established its own standard)

PDA
Personal Digital Assistant, e. g., Newton or Palm Pilot

Ping
Apple's social network for music

PDA
Personal Digital Assistant, z. B. Newton oder Palm

Ping
Website eines musikzentrierten sozialen Netzwerkes von Apple (seit 2010)

Podcast
Audio-Beitrag im Internet

PowerBook
Notebook von Apple

Retina Display
hohe Pixelauflösung beim iPod touch der 5. Generation

RIM®
Betriebssystem für die BlackBerry®-Smartphones

Screenlock
Software zum Schützen des Desktops mit einem Paßwort

Settings
Einstellungen

Shake to Shuffle
Zufallswiedergabe durch Schütteln des Gerätes für den iPod nano der 4. Generation

Shortcuts
Tastenkombination zum schnellern Aufrufen von Steuerbefehlen

Shuffle
ternärer (dreifacher) Rhythmus, vor allem im Blues und Jazz

Slider
Mobiltelephon mit Display und Tastatur, bei denen die obere und untere Gehäusehälfte gegeneinander verschiebbar sind

Smartphone
internetfähige Mobiltelephone

Terabyte
1000 Gigabyte, also eine Billion Byte, 10^{12}, 1000 000 000 000 Byte (griech: teras = Zeichen)

Touchphone
internetfähige Mobiltelephone ohne Tastatur

Touchscreen
berührungssensitives Glasdisplay

Trackpad
Steuerung für Programme von Desktop-Macs mit zwei Fingern. Dies erlaubt die Aktionen scrollen, vergrößern, verkleinern, Drehen von Bildern, schnelles Durchsuchen von Photosammlungen

Twitter
»to tweet = zwitschern«, Website für Kurznachrichten mit max. 140 Zeichen als soziales Netzwerk

UMTS
Universal Mobile Telecommunications System, Lizenzen für die Frequenzen von Mobiltelephonen

Virtual Reality
eine vom Computer simulierte Wirklichkeit, in die man sich mit Hilfe technischer Ausrüstungen scheinbar hineinversetzen kann. Dabei können alle sinnlichen Wahrnehmungen der realen Welt in Echtzeit als Analogon vermittelt werden

Walkman
mobiles Gerät für Audio-Kassetten von Sony (seit 1978)

Widget
Wi(ndow) (Ga)dget, eigenständig auf Ereignisse der Tastatur oder Maus reagierende Komponente eines graphischen Fenstersystems

WLAN
Wireless Local Area Network, drahtlose Übertragung für PC-Signale

Yahoo
Internetportal mit Nachrichtenportal, Suchfunktion und Mailprogramm

YouTube
Website für Videos

Zune
von Microsoft entwickelter Media-Player

Podcast
A digital audio contribution in the internet

PowerBook
Laptop computer by Apple

Retina Display
High pixel resolution of the 5th-generation iPod touch

RIM®
Operating system for the BlackBerry® smartphones

Screenlock
Software for restricting access to a computer with a password

Shake to Shuffle
Random playback by shaking device for the 4th generation iPod nano

Shortcuts
Links to frequently visited websites

Shuffle
Ternary (triple) rhythm, especially in blues and jazz

Slider
Mobile phone with touch screen and keypad

Smartphone
Internet-compatible mobile phones

Terabyte
1000 gigabytes, thus one trillion bytes, 10^{12}, 1 000 000 000 000 bytes (Greek: teras = sign)

Touchphone
Internet-compatible mobile phone without keypad

Touchscreen
Touch-sensitive glass display

Trackpad
Multitouch control for programs on desktop Macs with two fingers, allowing for scrolling, zooming in and out, rotating images, and the fast search of photo albums

Twitter
Name's origin: »to tweet,« website for short messages with max. 140 characters serving as a social network

UMTS
Universal Mobile Telecommunications System, licencies for the frequencies of mobile phones

Virtual reality
Computer-generated simulated reality which users can enter via technological equipment. All sensual perceptions of the real world can be communicated in real-time as an analogon.

Walkman
Mobile device from Sony for playing audio cassette tapes (since 1978)

Widget
Wi(ndow) (Ga)dget, a reusable element of a graphical user interface that displays information and independently reacts to events on the keyboard or mouse

WLAN
Wireless Local Area Network, for linking two or more devices using a wireless distribution method

Yahoo
A web portal with news and mail function and search engine

YouTube
Website for video sharing

Zune
Software by Microsoft for organizing, playing and downloading music tracks

Literatur

Aicher, Otl, und Robert Kuhn: *Greifen und Griffe*, Köln 1987, darin: »Greifen und Begreifen«, S. 8 ff.

Anon.: »Als der Computer sexy wurde«, in: *Frankfurter Neue Presse*, 24. Jan. 2009, S. 5.

Anon.: »Architektur. iPod-Hochhaus in Dubai«, in: *Der Spiegel*, Nr. 2, 2007.

Barthes, Roland: *Mythen des Alltags*, Frankfurt am Main 1964 (Paris 1957).

Beyer, Susanne: »Ich bin dann mal off. Über die Kunst des Müßiggangs im digitalen Zeitalter«, in: *Der Spiegel*, Nr. 28, 2010, S. 56–67.

Brauck, Markus, Isabell Hülsen und Marcel Rodenbach: »Steve sieht alles«, in: *Der Spiegel*, Nr. 26, 2010, S. 120–122.

Brinkbäumer, Klaus und Thomas Schulz: »Der iKult. Wie Apple die Welt verführt«, in: *Der Spiegel*, Nr. 17, 2010, S. 66–78.

Bürdek, Bernhard E.: *Der Apple Macinthosh*, Frankfurt am Main 1997 (Reihe *Design-Klassiker*, Nr. 9, hrsg. von Volker Fischer).

Erdmann, Gerald und Charlotte Stanek: *iPod und iTunes*, Beijing und Köln 2007.

Esslinger, Hartmut: *Schwungrat*, Weinheim 2009.

Feingold, Howard: *Virtual Reality*, New York 1991.

Fischer, Volker: »Apple Macinthosh Frogdesign Hartmut Esslinger«, in : Albus, Volker et al.: *Design! Das 20. Jahrhundert*, München, London und New York 2000, S. 160 f.; http://de.wikipedia.org/wiki/Apple iMac.

Fischer, Volker: *Design heute. Maßstäbe. Formgebung zwischen Industrie und Kunst-Stück*, München 1988.

Fischer, Volker: »Emotionen in der Digitale. Eine Phänomenologie elektronischer ›devices‹«, in: Bürdek, Bernhard E.: *Der digitale Wahn*, Frankfurt am Main 2001, S. 44–64.

Gabor, Simon, und Felix Wadewitz: »Microsoft sucht Anschluß. Konzern will mit einem Smartphone für junge Leute Apple und Google angreifen«, in: *Frankfurter Rundschau*, 7. Apr. 2010, S. 17.

Garz, Joachim: *Die Apple-Story: Aufstieg, Niedergang und »Wieder-Auferstehung« des Unternehmens rund um Steve Jobs*, Kilchberg 2005.

Gros, Jochen: »Neue Bilderschrift« sowie Volker Fischer: »Die diskursive Logik des Präsentativen, Anmerkungen zum ›Projekt Bilderschrift‹«, in: Jochen Gros, *Digital Fiktional*, Frankfurt am Main 1993.

Haas, Daniel: »Ein Schatten seiner selbst«, in: *Frankfurter Allgemeine Zeitung*, 11. Dez. 2010, S. Z4 (Essay über Karl Lagerfeld).

Haring, Bruce: *MP3. Die digitale Revolution in der Musikindustrie*, Freiburg 2002.

Hirzel, J., M. Franke, M. Kietzmann und A. Kusitzky: »Der große Imperator. Wie Steve Jobs mit Apple die Welt revolutioniert und bevormundet«, in: *Focus*, Nr. 35, 2010, S. 124–130.

http://de.wikipedia.org.wiki/Apple sowie Apple Inc: *Report on iPod Manufacturing*, Aug. 2006.

http://de.wikipedia.org/wiki/Apple.

http://de.wikipedia.org.wiki/iMac.

http://de.wikipedia.org.wiki/iPhone.

http://de.wikipedia.org.wiki/iPod.

Hülsen, Isabell, und Martin U. Müller: »Angriff der Trittpadfahrer«, in: *Der Spiegel*, Nr. 26, 2010, S. 122.

Koch, Christoph: *Ich bin dann mal offline. Ein Selbstversuch. Leben ohne Handy und Internet*, München 2010.

Kunkel, Paul: *AppleDesign. The work of the Apple Industrial Design Group*, New York 1997.

Levy, Steven: *The Perfect Thing. How the iPod Shuffles Commerce, Culture, and Coolness*, New York 2007.

Linzmayer, Owen W.: *Apple streng vertraulich. Die Tops und Flops der Macintosh-Geschichte*, Zürich 2000.

Matting, Matthias: »Die Tablet-Welle«, in: *Focus*, Nr. 35, 2010, S. 78 f.

Mejias, Jordan: »Die Welt leuchtet auf. Der iPad ist da«, in: *Frankfurter Allgemeine Zeitung*, 6. Apr. 2010.

Molloy, Andy: »Apple Computer Reading List, bn (= burn notice)«, in: *Network Forum*, 23. Okt. 2001.

Notomi, Yasukuni: *iPod – Das Buch zum Kult-Player*, Köln 2005.

Oehmke, Philipp, und Tobias Rapp: »Ein Jahrzehnt für die Ewigkeit (50 Jahre Beatles)«, in: *Stern*, Nr. 21, 2010, S. 109–121, S. 119 f.

Prechel, Anja: »Frankfurt im iPad-Fieber«, in: *Frankfurter Neue Presse*, 29. Mai 2010, S. 6.

Rühle, Alex: *Ohne Netz*, Köln 2010.

Bibliography

Aicher, Otl, and Robert Kuhn: *Greifen und Griffe*, Köln 1987, therein: »Greifen und Begreifen«, pp. 8 ff.

Anon.: »Als der Computer sexy wurde«, in: *Frankfurter Neue Presse*, 24 Jan. 2009, p. 5.

Anon.: »Architektur. iPod-Hochhaus in Dubai«, in: *Der Spiegel*, no. 2, 2007.

Barthes, Roland: *Mythen des Alltags*, Frankfurt am Main 1964 (Paris 1957).

Beyer, Susanne: »Ich bin dann mal off. Über die Kunst des Müßiggangs im digitalen Zeitalter«, in: *Der Spiegel*, no. 28, 2010, pp. 56–67.

Brauck, Markus, Isabell Hülsen, and Marcel Rodenbach: »Steve sieht alles«, in: *Der Spiegel*, no. 26, 2010, pp. 120–122.

Brinkbäumer, Klaus, and Thomas Schulz: »Der iKult. Wie Apple die Welt verführt«, in: *Der Spiegel*, no. 17, 2010, pp. 66–78.

Bürdek, Bernhard E.: *Der Apple Macinthosh*, Frankfurt am Main 1997 (series *Design-Klassiker*, no. 9, ed. by Volker Fischer).

Erdmann, Gerald, and Charlotte Stanek: *iPod und iTunes*, Beijing and Cologne 2007.

Esslinger, Hartmut: *Schwungrat*, Weinheim 2009.

Feingold, Howard: *Virtual Reality*, New York 1991.

Fischer, Volker: »Apple Macinthosh Frogdesign Hartmut Esslinger«, in : Albus, Volker et al.: *Design! Das 20. Jahrhundert*, München, London und New York 2000, pp. 160 f.; http://de.wikipedia.org/wiki/Apple iMac.

Fischer, Volker: *Design heute. Maßstäbe. Formgebung zwischen Industrie und Kunst-Stück*, Munich 1988.

Fischer, Volker: »Emotionen in der Digitale. Eine Phänomenologie elektronischer ›devices‹«, in: Bürdek, Bernhard E.: *Der digitale Wahn*, Frankfurt am Main 2001, pp. 44–64.

Gabor, Simon, und Felix Wadewitz: »Microsoft sucht Anschluss. Konzern will mit einem Smartphone für junge Leute Apple und Google angreifen«, in: *Frankfurter Rundschau*, 7 Apr. 2010, p. 17.

Garz, Joachim: *Die Apple-Story: Aufstieg, Niedergang und »Wieder-Auferstehung« des Unternehmens rund um Steve Jobs*, Kilchberg 2005.

Gros, Jochen: »Neue Bilderschrift« and Volker Fischer: »Die diskursive Logik des Präsentativen, Anmerkungen zum ›Projekt Bilderschrift‹« in: Jochen Gros, *Digital Fiktional*, Frankfurt am Main 1993.

Haas, Daniel: »Ein Schatten seiner selbst«, in: *Frankfurter Allgemeine Zeitung*, 11 Dec. 2010, S.Z4 (essay about Karl Lagerfeld).

Haring, Bruce: *MP3. Die digitale Revolution in der Musikindustrie*, Freiburg 2002.

Hirzel, J., M. Franke, M. Kietzmann and A. Kusitzky: »Der große Imperator. Wie Steve Jobs mit Apple die Welt revolutioniert und bevormundet«, in: *Focus*, no. 35, 2010, pp. 124–130.

http://de.wikipedia.org.wiki/Apple sowie Apple Inc: *Report on iPod Manufacturing*, Aug. 2006.

http://de.wikipedia.org/wiki/Apple.

http://de.wikipedia.org.wiki/iMac.

http://de.wikipedia.org.wiki/iPhone.

http://de.wikipedia.org.wiki/iPod.

Hülsen, Isabell, and Martin U. Müller: »Angriff der Trittpadfahrer«, in: *Der Spiegel*, no. 26, 2010, p. 122.

Koch, Christoph: *Ich bin dann mal offline. Ein Selbstversuch. Leben ohne Handy und Internet*, Munich 2010.

Kunkel, Paul: *AppleDesign. The work of the Apple Industrial Design Group*, New York 1997.

Levy, Steven: *The Perfect Thing. How the iPod Shuffles Commerce, Culture, and Coolness*, New York 2007.

Linzmayer, Owen W.: *Apple streng vertraulich. Die Tops und Flops der Macintosh-Geschichte*, Zurich 2000.

Matting, Matthias: »Die Tablet-Welle«, in: *Focus*, no. 35, 2010, pp. 78 f.

Mejias, Jordan: »Die Welt leuchtet auf. Der iPad ist da«, in: *Frankfurter Allgemeine Zeitung*, 6. Apr. 2010.

Molloy, Andy: »Apple Computer Reading List, bn (= burn notice)«, in: *Network Forum*, 23 Oct. 2001.

Notomi, Yasukuni: *iPod – Das Buch zum Kult-Player*, Cologne 2005.

Oehmke, Philipp, and Tobias Rapp: »Ein Jahrzehnt für die Ewigkeit (50 Jahre Beatles)«, in: *Stern*, no. 21, 2010, pp. 109–121, p. 119 f.

Prechel, Anja: »Frankfurt im iPad-Fieber«, in: *Frankfurter Neue Presse*, 29 Mai. 2010, p. 6.

Rühle, Alex: *Ohne Netz*, Cologne 2010.

Schirrmacher, Frank: *Payback*, München 2009.

Schönert, Ulf: »Das neue Puschenkino«, in: *Stern*, Nr. 50, 2010, S. 83–90.

Sloterdijk, Peter: »Welt-Ortsgespräche«, in: Lamprecht, Rudi: *Zukunft Mobile Kommunikation*, Frankfurt am Main 2001, S. 193–244, S. 201 f.

Spehr, Michael: »Logitech Audio Station für den iPod. Die günstige Docking-Station«, in: *Frankfurter Allgemeine Zeitung*, 16. Jan. 2007.

Spehr, Michael: »Schicke Schale und schöne Schnappschüsse (iPhone 4)«, in: *Frankfurter Allgemeine Zeitung*, 29. Jun. 2010, S. T6.

Spehr, Michael: »Forscher Wettstreit der flachen Webpads«, in: *Frankfurter Allgemeine Sonntagszeitung*, 5. Sep. 2010, S. V10.

Steffen, Dagmar: *Design als Produktsprache. Der »Offenbacher Ansatz« Theorie und Praxis*, Mit Beiträgen von Bernhard E. Bürdek, Volker Fischer und Jochen Gros, Frankfurt am Main 2000.

Thadeusz, Frank: »Psychologie. Drang zum Ding. Experten rätseln über die bizarre sexuelle Spielart der Objektophilie«, in: *Der Spiegel*, Nr. 19, 2007, S. 160.

Wagner, Thomas: »Berühre die Welt. Mit dem ultimativen Touch: Apples neues iPhone«, in: *Frankfurter Allgemeine Zeitung*, 11. Jan. 2007.

Weiß, Matthias: »Mobiles Stillleben. Revolutioniert das Smartphone die Produktionsweise der Künstler?«, in: *Kunstzeitung*, Nr. 1, 2011, S. 1.

Wozniak, Steve und Gina Smith: *iWoz. Wie ich den Personal Computer erfand und Apple mitgründete*, München 2006.

Young, Jeffrey S. und William L. Simon: *Steve Jobs und die Geschichte eines außergewöhnlichen Unternehmens*, Frankfurt am Main 2007.

Schirrmacher, Frank: *Payback*, Munich 2009.

Schönert, Ulf: »Das neue Puschenkino«, in: *Stern*, no. 50, 2010, pp. 83–90.

Sloterdijk, Peter: »Welt-Ortsgespräche«, in: Lamprecht, Rudi: *Zukunft Mobile Kommunikation*, Frankfurt am Main 2001, pp. 193–244, pp. 201 f.

Spehr, Michael: »Logitech Audio Station für den iPod. Die günstige Docking-Station«, in: *Frankfurter Allgemeine Zeitung*, 16 Jan. 2007.

Spehr, Michael: »Schicke Schale und schöne Schnappschüsse (iPhone 4)«, in: *Frankfurter Allgemeine Zeitung*, 29 Jun. 2010, p. T6.

Spehr, Michael: »Forscher Wettstreit der flachen Webpads«, in: *Frankfurter Allgemeine Sonntagszeitung*, 5 Sep. 2010, p. V10.

Steffen, Dagmar: *Design als Produktsprache. Der »Offenbacher Ansatz« Theorie und Praxis*, with contributionns by Bernhard E. Bürdek, Volker Fischer, and Jochen Gros, Frankfurt am Main 2000.

Thadeusz, Frank: »Psychologie. Drang zum Ding. Experten rätseln über die bizarre sexuelle Spielart der Objektophilie«, in: *Der Spiegel*, no. 19, 2007, p. 160.

Wagner, Thomas: »Berühre die Welt. Mit dem ultimativen Touch: Apples neues iPhone«, in: *Frankfurter Allgemeine Zeitung*, 11 Jan. 2007.

Weiß, Matthias: »Mobiles Stillleben. Revolutioniert das Smartphone die Produktionsweise der Künstler?«, in: *Kunstzeitung*, no. 1, 2011, p. 1.

Wozniak, Steve, and Gina Smith: *iWoz. Wie ich den Personal Computer erfand und Apple mitgründete*, Munich 2006.

Young, Jeffrey S., and William L. Simon: *Steve Jobs und die Geschichte eines außergewöhnlichen Unternehmens*, Frankfurt am Main 2007.

Bildnachweis / Credits

1&1 Internet AG 203
4tiitoo 196
Acer 207
Altec Lansing 102, 116, 211
Amazon 190
Ap images 11
Apple cover photo, 47, 51, 53, 54, 55, 56, 58, 62, 63, 64, 65, 66, 68, 69, 70, 71, 94, 154, 193, 216, 217, 218, 219, 220, 221, 222, 227, 232, 233, 251,
Apple www.apple.com/mac/app-store 225, 226
Apronti 34
Archos 79, 93, 200
Arts in the city 112
Asus 197
Audi 242, 243
Avox 107
Bang & Olufsen 126
Barnes & Noble 188
Batoul Apps 234, 235
Beatles Museum Halle 8, 9, 10
Belkin 160
Bergmann 120, 132
BlackBerry® 179, 180
BMW 241
Bowers & Wilkins 114
Brabus 246, 247, 248
Cisco 209
Creative 83, 204
DeinDesign 158
Dell 206
Dettmar, Uwe 1, 2, 3, 4, 5, 6, 7, 13, 15, 17, 19, 20, 21, 22, 24, 30, 31, 32, 35, 36, 37, 38, 40, 41, 42, 43, 45, 48, 49, 50, 59, 60, 95, 96, 139, 148, 163, 192
Deutsches Patent- und Markenamt 76
Digital Group Audio 103
Düvel, Sebastian 67
Edifier 113
Elastica 142
Elecom 74
Fader, John /wikipedia.de 25
Finite Elemente 124
Focus 195
Frankie's Garage 141
Freiwild 140
frog design 39, 212, 213, 214, 215, 238, 239, 240
Garmin 245
Geneva 110
Getty images 12, 150
Golfoholic 75
Google 168
Grassi-Museum für Angewandte Kunst Leipzig 61
Grove Made 157
Grundig 23, 81, 91
hama 257, 258
Hanvon 189, 198
Hottrix 230
HTC 166, 169
IF International Forum Design Frontispiez, 44, 46, 52, 57, 72, 111, 149
Iriver 86, 183
James Law Cybertecture 144, 145

JBL 100, 115, 127
Junaio 236, 237
Jung 147, 254, 255
Kardon Harman 104
Kassan, David Jon 259–262
Kenwood 182, 244
Koziol 143
Kunstflug 263
Lapàporter 156
Lenco 88, 97, 101, 118, 128, 256
Lenovo 208
LG 167, 177
Lürzers Archiv 223, 224, 228, 229
Microsoft 84, 85
Morgan, Peter J., 2010 / Apex Imageworks 153
Motorola 170, 176, 181
Msi Computers 205
Native Union 122
Nike 252, 253
Nintendo 26, 27, 28, 29
Nokia 171, 175, 178
Odys 184
Pack and smooch 155
Palm 164, 165
Panasonic 33
Parrot 109
Philips 16, 18, 77, 89, 210
Pioneer 98, 99
re:store 231
Rotaliana 133
Samsung 78, 80, 87, 90, 172, 174, 199
SanDisk 82, 92
Sharp 119
Silver Seiko 134
Sonoro 121, 129
Sonos 117
Sony 14, 105, 123, 185
Sony Ericsson 173
Soulra 108
Der Spiegel 194
switcheasy 73, 135, 136, 137, 138, 159, 161, 162
Tchibo 125
Technabob 152
Thalia Buchhandelsgruppe 186
Tivoli 106, 131
Toshiba 202,
TrekStor 130
Twelve South 191
Urban Tool 249, 250
ViewSonic 187, 201
Xayni 151
Zumtobel 146

BlackBerry®, RIM®, Research In Motion® and related trademarks, names and logos are the property of Research In Motion Limited and are registered and/or used in the U.S. and countries around the world. Used under license from Research In Motion Limited.